Guojie Li

# Information Science & Technology in China: A Roadmap to 2050

Chinese Academy of Sciences

Guojie Li
*Editor*

# Information Science & Technology in China: A Roadmap to 2050

With 15 figures

Science Press
Beijing

Springer

*Editor*
Guojie Li
Institute of Computing Technology
Chinese Academy of Sciences
100190, Beijing, China
E-mail: lig@ict.ac.cn

ISBN 978-7-03-029811-9
Science Press Beijing

ISBN 978-3-642-19070-4 e-ISBN 978-3-642-19071-1
Springer Heidelberg Dordrecht London New York



*Cover design*: Frido Steinen-Broo, EStudio Calamar, Spain

Printed on acid-free paper

Springer is part of Springer Science+Business Media (www.springer.com)

# Members of the Editorial Committee and the Editorial Office

## Editor-in-Chief

Yongxiang Lu

## Editorial Committee

Yongxiang Lu    Chunli Bai    Erwei Shi    Xin Fang    Zhigang Li    Xiaoye Cao    Jiaofeng Pan

# Research Group on Information Science & Technology of the Chinese Academy of Sciences

| | | |
|---|---|---|
| **Head:** | Guojie Li | Institute of Computing Technology, Chinese Academy of Sciences |
| **Members:** | Houzhi Zheng | Institute of Semiconductors, Chinese Academy of Sciences |
| | Ziqiang Hou | Institute of Acoustics, Chinese Academy of Sciences |
| | Zhiwei Xu | Institute of Computing Technology, Chinese Academy of Sciences |
| | Dexin Wu | Institute of Microelectronics, Chinese Academy of Sciences |
| | Huiming Lin | Institute of Software of Chinese Academy of Sciences |
| | Guangcan Guo | University of Science and Technology of China |
| | Runsheng Chen | Institute of Biophysics, Chinese Academy of Sciences |
| | Songlin Feng | Institute of Micro-system and Information Technology, Chinese Academy of Sciences |
| | Feiyue Wang | Institute of Automation, Chinese Academy of Sciences |
| | Zhongzhi Shi | Institute of Computing Technology, Chinese Academy of Sciences |
| | Zhiyong Liu | Institute of Computing Technology, Chinese Academy of Sciences |
| | Xuehai Hong | Institute of Computing Technology, Chinese Academy of Sciences |

# Foreword to the Roadmaps 2050*

China's modernization is viewed as a transformative revolution in the human history of modernization. As such, the Chinese Academy of Sciences (CAS) decided to give higher priority to the research on the science and technology (S&T) roadmap for priority areas in China's modernization process. What is the purpose? And why is it? Is it a must? I think those are substantial and significant questions to start things forward.

## Significance of the Research on China's S&T Roadmap to 2050

We are aware that the National Mid- and Long-term S&T Plan to 2020 has already been formed after two years' hard work by a panel of over 2000 experts and scholars brought together from all over China, chaired by Premier Wen Jiabao. This clearly shows that China has already had its S&T blueprint to 2020. Then, why did CAS conduct this research on China's S&T roadmap to 2050?

In the summer of 2007 when CAS was working out its future strategic priorities for S&T development, it realized that some issues, such as energy, must be addressed with a long-term view. As a matter of fact, some strategic researches have been conducted, over the last 15 years, on energy, but mainly on how to best use of coal, how to best exploit both domestic and international oil and gas resources, and how to develop nuclear energy in a discreet way. Renewable energy was, of course, included but only as a supplementary energy. It was not yet thought as a supporting leg for future energy development. However, greenhouse gas emissions are becoming a major world concern over

* It is adapted from a speech by President Yongxiang Lu at the First High-level Workshop on China's S&T Roadmap for Priority Areas to 2050, organized by the Chinese Academy of Sciences, in October, 2007.

the years, and how to address the global climate change has been on the agenda. In fact, what is really behind is the concern for energy structure, which makes us realize that fossil energy must be used cleanly and efficiently in order to reduce its impact on the environment. However, fossil energy is, pessimistically speaking, expected to be used up within about 100 years, or optimistically speaking, within about 200 years. Oil and gas resources may be among the first to be exhausted, and then coal resources follow. When this happens, human beings will have to refer to renewable energy as its major energy, while nuclear energy as a supplementary one. Under this situation, governments of the world are taking preparatory efforts in this regard, with Europe taking the lead and the USA shifting to take a more positive attitude, as evidenced in that: while fossil energy has been taken the best use of, renewable energy has been greatly developed, and the R&D of advanced nuclear energy has been reinforced with the objective of being eventually transformed into renewable energy. The process may last 50 to 100 years or so. Hence, many S&T problems may come around. In the field of basic research, for example, research will be conducted by physicists, chemists and biologists on the new generation of photovoltaic cell, dye-sensitized solar cells (DSC), high-efficient photochemical catalysis and storage, and efficient photosynthetic species, or high-efficient photosynthetic species produced by gene engineering which are free from land and water demands compared with food and oil crops, and can be grown on hillside, saline lands and semi-arid places, producing the energy that fits humanity. In the meantime, although the existing energy system is comparatively stable, future energy structure is likely to change into an unstable system. Presumably, dispersive energy system as well as higher-efficient direct current transmission and storage technology will be developed, so will be the safe and reliable control of network, and the capture, storage, transfer and use of $CO_2$, all of which involve S&T problems in almost all scientific disciplines. Therefore, it is natural that energy problems may bring out both basic and applied research, and may eventually lead to comprehensive structural changes. And this may last for 50 to 100 years or so. Taking the nuclear energy as an example, it usually takes about 20 years or more from its initial plan to key technology breakthroughs, so does the subsequent massive application and commercialization. If we lose the opportunity to make foresighted arrangements, we will be lagging far behind in the future. France has already worked out the roadmap to 2040 and 2050 respectively for the development of the 3rd and 4th generation of nuclear fission reactors, while China has not yet taken any serious actions. Under this circumstance, it is now time for CAS to take the issue seriously, for the sake of national interests, and to start conducting a foresighted research in this regard.

This strategic research covers over some dozens of areas with a long-term view. Taking agriculture as an example, our concern used to be limited only to the increased production of high-quality food grains and agricultural by-products. However, in the future, the main concern will definitely be given to the water-saving and ecological agriculture. As China is vast in territory, diversified technologies in this regard are the appropriate solutions. Animal husbandry has been used by developed countries, such as Japan and Denmark, to make bioreactor and pesticide as well. Plants have been used by Japan to make bioreactors which are safer and cost-effective than that made from animals. Potato, strawberry, tomato and the like have been bred in germ-free greenhouses, and value-added products have been made through gene transplantation technology. Agriculture in China must not only address the food demands from its one billions-plus population, but also take into consideration of the value-added agriculture by-products and the high-tech development of agriculture as well. Agriculture in the future is expected to bring out some energies and fuels needed by both industry and man's livelihood as well. Some developed countries have taken an earlier start to conduct foresighted research in this regard, while we have not yet taken sufficient consideration.

Population is another problem. It will be most likely that China's population will not drop to about 1 billion until the end of this century, given that the past mistakes of China's population policy be rectified. But the subsequent problem of ageing could only be sorted out until the next century. The current population and health policies face many challenges, such as, how to ensure that the 1.3 to 1.5 billion people enjoy fair and basic public healthcare; the necessity to develop advanced and public healthcare and treatment technologies; and the change of research priority to chronic diseases from infectious diseases, as developed countries have already started research in this regard under the increasing social and environmental change. There are many such research problems yet to be sorted out by starting from the basic research, and subsequent policies within the next 50 years are in need to be worked out.

Space and oceans provide humanity with important resources for future development. In terms of space research, the well-known Manned Spacecraft Program and China's Lunar Exploration Program will last for 20 or 25 years. But what will be the whole plan for China's space technology? What is the objective? Will it just follow the suit of developed countries? It is worth doing serious study in this regard. The present spacecraft is mainly sent into space with chemical fuel propellant rocket. Will this traditional propellant still be used in future deep space exploration? Or other new technologies such as electrical propellant, nuclear energy propellant, and solar sail technologies be developed? We haven't yet done any strategic research over these issues, not even worked out any plans. The ocean is abundant in mineral resources, oil and gas, natural gas hydrate, biological resources, energy and photo-free biological evolution, which may arise our scientific interests. At present, many countries have worked out new strategic marine plans. Russia, Canada, the USA, Sweden and Norway have centered their contention upon the North Pole, an area of strategic significance. For this, however, we have only limited plans.

The national and public security develops with time, and covers both conventional and non-conventional security. Conventional security threats only refer to foreign invasion and warfare, while, the present security threat may come out from any of the natural, man-made, external, interior, ecological, environmental, and the emerging networking (including both real and virtual) factors. The conflicts out of these must be analyzed from the perspective of human civilization, and be sorted out in a scientific manner. Efforts must be made to root out the cause of the threats, while human life must be treasured at any time.

In general, it is necessary to conduct this strategic research in view of the future development of China and mankind as well. The past 250 years' industrialization has resulted in the modernization and better-off life of less than 1 billion people, predominantly in Europe, North America, Japan and Singapore. The next 50 years' modernization drive will definitely lead to a better-off life for 2–3 billion people, including over 1 billion Chinese, doubling or tripling the economic increase over that of the past 250 years, which will, on the one hand, bring vigor and vitality to the world, and, on the other hand, inevitably challenge the limited resources and eco-environment on the earth. New development mode must be shaped so that everyone on the earth will be able to enjoy fairly the achievements of modern civilization. Achieving this requires us, in the process of China's modernization, to have a foresighted overview on the future development of world science and human civilization, and on how science and technology could serve the modernization drive. S&T roadmap for priority areas to 2050 must be worked out, and solutions to core science problems and key technology problems must be straightened out, which will eventually provide consultations for the nation's S&T decision-making.

### Possibility of Working out China's S&T Roadmap to 2050

Some people held the view that science is hard to be predicted as it happens unexpectedly and mainly comes out of scientists' innovative thinking, while, technology might be predicted but at the maximum of 15 years. In my view, however, S&T foresight in some areas seems feasible. For instance, with the exhaustion of fossil energy, some smart people may think of transforming solar energy into energy-intensive biomass through improved high-efficient solar thin-film materials and devices, or even developing new substitute. As is driven by huge demands, many investments will go to this emerging area. It is, therefore, able to predict that, in the next 50 years, some breakthroughs will undoubtedly be made in the areas of renewable energy and nuclear energy as well. In terms of solar energy, for example, the improvement of photoelectric conversion efficiency and photothermal conversion efficiency will be the focus. Of course, the concrete technological solutions may be varied, for example, by changing the morphology of the surface of solar cells and through the reflection, the entire spectrum can be absorbed more efficiently; by developing multi-layer functional thin-films for transmission and absorption; or by introducing of nanotechnology and quantum control technology, etc. Quantum control research used to limit mainly to the solution to information functional materials. This is surely too narrow. In the future, this research is expected to be extended to the energy issue or energy-based basic research in cutting-edge areas.

In terms of computing science, we must be confident to forecast its future development instead of simply following suit as we used to. This is a possibility rather than wild fancies. Information scientists, physicists and biologists could be engaged in the forward-looking research. In 2007, the Nobel Physics Prize was awarded to the discovery of colossal magneto-resistance, which was, however, made some 20 years ago. Today, this technology has already been applied to hard disk store. Our conclusion made, at this stage, is that: it is possible to make long-term and unconventional S&T predictions, and so is it to work out China's S&T roadmap in view of long-term strategies, for example, by 2020 as the first step, by 2030 or 2035 as the second step, and by 2050 as the maximum.

This possibility may also apply to other areas of research. The point is to emancipate the mind and respect objective laws rather than indulging in wild fancies. We attribute our success today to the guidelines of emancipating the mind and seeking the truth from the facts set by the Third Plenary Session of the 11th Central Committee of the Communist Party of China in 1979. We must break the conventional barriers and find a way of development fitting into China's reality. The history of science tells us that discoveries and breakthroughs could only be made when you open up your mind, break the conventional barriers, and make foresighted plans. Top-down guidance on research with increased financial support and involvement of a wider range of talented scientists is not in conflict with demand-driven research and free discovery of science as well.

### Necessity of CAS Research on China's S&T Roadmap to 2050

Why does CAS launch this research? As is known, CAS is the nation's highest academic institution in natural sciences. It targets at making basic, forward-looking and strategic research and playing a leading role in China's science. As such, how can it achieve this if without a foresighted view on science and technology? From the perspective of CAS, it is obligatory to think, with a global view, about what to do after the 3rd Phase of the Knowledge Innovation Program (KIP). Shall we follow the way as it used to? Or shall we, with a view of national interests, present our in-depth insights into different research disciplines, and make efforts to reform the organizational structure and system, so that the innovation capability of CAS and the nation's science and technology mission will be raised to a new height? Clearly, the latter is more positive. World science and technology develops at a lightening speed. As global economy grows, we are aware that we will be lagging far behind if without making progress, and will lose the opportunity if without making foresighted plans. S&T innovation requires us to make joint efforts, break the conventional barriers and emancipate the mind. This is also what we need for further development.

The roadmap must be targeted at the national level so that the strategic research reports will form an important part of the national long-term program. CAS may not be able to fulfill all the objectives in the reports. However, it can select what is able to do and make foresighted plans, which will eventually help shape the post-2010 research priorities of CAS and the guidelines for its future reform.

Once the long-term roadmap and its objectives are identified, system mechanism, human resources, funding and allocation should be ensured for full implementation. We will make further studies to figure out: What will happen to world innovation system within the next 30 to 50 years? Will universities, research institutions and enterprises still be included in the system? Will research institutes become grid structure? When the cutting-edge research combines basic science and high-tech and the transformative research integrates the cutting-edge research with industrialization, will that be the research trend in some disciplines? What will be the changes for personnel structure, motivation mechanism and upgrading mechanism within the innovation system? Will there be any changes for the input and structure of innovation resources? If we could have a clear mind of all the questions, make foresighted plans and then dare to try out in relevant CAS institutes, we will be able to pave a way for a more competitive and smooth development.

Social changes are without limit, so are the development of science and technology, and innovation system and management as well. CAS must keep moving ahead to make foresighted plans not only for science and technology, but also for its organizational structure, human resources, management modes, and resource structures. By doing so, CAS will keep standing at the forefront of science and playing a leading role in the national innovation system, and even, frankly speaking, taking the lead in some research disciplines in the world. This is, in fact, our purpose of conducting the strategic research on China's S&T roadmap.

Prof. Dr.-Ing. Yongxiang Lu

President of the Chinese Academy of Sciences

# Preface to the Roadmaps 2050

CAS is the nation's think tank for science. Its major responsibility is to provide S&T consultations for the nation's decision-makings and to take the lead in the nation's S&T development.

In July, 2007, President Yongxiang Lu made the following remarks: "In order to carry out the Scientific Outlook of Development through innovation, further strategic research should be done to lay out a S&T roadmap for the next 20–30 years and key S&T innovation disciplines. And relevant workshops should be organized with the participation of scientists both within CAS and outside to further discuss the research priorities and objectives. We should no longer confine ourselves to the free discovery of science, the quantity and quality of scientific papers, nor should we satisfy ourselves simply with the Principal Investigators system of research. Research should be conducted to address the needs of both the nation and society, in particular, the continued growth of economy and national competitiveness, the development of social harmony, and the sustainability between man and nature."

According to the Executive Management Committee of CAS in July, 2007, CAS strategic research on S&T roadmap for future development should be conducted to orchestrate the needs of both the nation and society, and target at the three objectives: the growth of economy and national competitiveness, the development of social harmony, and the sustainability between man and nature.

In August, 2007, President Yongxiang Lu further put it: "Strategic research requires a forward-looking view over the world, China, and science & technology in 2050. Firstly, in terms of the world in 2050, we should be able to study the perspectives of economy, society, national security, eco-environment, and science & technology, specifically in such scientific disciplines as energy, resources, population, health, information, security, eco-environment, space and oceans. And we should be aware of where the opportunities and challenges lie. Secondly, in terms of China's economy and society in 2050, we should take into consideration of factors like: objectives, methods, and scientific supports needed for economic structure, social development, energy structure, population and health, eco-environment, national security and innovation capability. Thirdly, in terms of the guidance of Scientific Outlook of Development on science and technology, it emphasizes the people's interests and development, science and technology, science and economy, science and society, science and eco-environment, science and culture, innovation and collaborative development. Fourthly, in terms of the supporting role of research in scientific development, this includes how to optimize the economic structure and boost economy, agricultural development, energy structure, resource conservation, recycling economy, knowledge-based society, harmonious coexistence between man and nature, balance of regional development, social harmony, national security, and international cooperation. Based on these, the role of CAS will be further identified."

Subsequently, CAS launched its strategic research on the roadmap for priority areas to 2050, which comes into eighteen categories including: energy, water resources, mineral resources, marine resources, oil and gas, population and health, agriculture, eco-environment, biomass resources, regional development, space, information, advanced manufacturing, advanced materials, nano-science, big science facilities, cross-disciplinary and frontier research, and national and public security. Over 300 CAS experts in science, technology, management and documentation & information, including about 60 CAS members, from over 80 CAS institutes joined this research.

Over one year's hard work, substantial progress has been made in each research group of the scientific disciplines. The strategic demands on priority areas in China's modernization drive to 2050 have been strengthened out; some core science problems and key technology problems been set forth; a relevant S&T roadmap been worked out based on China's reality; and eventually the strategic reports on China's S&T roadmap for eighteen priority areas to 2050 been formed. Under the circumstance, both the Editorial Committee and Writing Group, chaired by President Yongxiang Lu, have finalized the general report. The research reports are to be published in the form of CAS strategic research serial reports, entitled *Science and Technology Roadmap to China 2050: Strategic Reports of the Chinese Academy of Sciences*.

The unique feature of this strategic research is its use of S&T roadmap approach. S&T roadmap differs from the commonly used planning and technology foresight in that it includes science and technology needed for the future, the roadmap to reach the objectives, description of environmental changes, research needs, technology trends, and innovation and technology development. Scientific planning in the form of roadmap will have a clearer scientific objective, form closer links with the market, projects selected be more interactive and systematic, the solutions to the objective be defined, and the plan be more feasible. In addition, by drawing from both the foreign experience on roadmap research and domestic experience on strategic planning, we have formed our own ways of making S&T roadmap in priority areas as follows:

**(1) Establishment of organization mechanism for strategic research on S&T roadmap for priority areas**

The Editorial Committee is set up with the head of President Yongxiang Lu and the involvement of Chunli Bai, Erwei Shi, Xin Fang, Zhigang Li, Xiaoye Cao and Jiaofeng Pan. And the Writing Group was organized to take responsibility of the research and writing of the general report. CAS Bureau of Planning and Strategy, as the executive unit, coordinates the research, selects the scholars, identifies concrete steps and task requirements, sets forth research approaches, and organizes workshops and independent peer reviews of the research, in order to ensure the smooth progress of the strategic research on the S&T roadmap for priority areas.

**(2) Setting up principles for the S&T roadmap for priority areas**

The framework of roadmap research should be targeted at the national level, and divided into three steps as immediate-term (by 2020), mid-term (by 2030) and long-term (by 2050). It should cover the description of job requirements, objectives, specific tasks, research approaches, and highlight core science problems and key technology problems, which must be, in general, directional, strategic and feasible.

**(3) Selection of expertise for strategic research on the S&T roadmap**

Scholars in science policy, management, information and documentation, and chief scientists of the middle-aged and the young should be selected to form a special research group. The head of the group should be an outstanding scientist with a strategic vision, strong sense of responsibility and coordinative capability. In order to steer the research direction, chief scientists should be selected as the core members of the group to ensure that the strategic research in priority areas be based on the cutting-edge and frontier research. Information and documentation scholars should be engaged in each research group to guarantee the efficiency and systematization of the research through data collection and analysis. Science policy scholars should focus on the strategic demands and their feasibility.

**(4) Organization of regular workshops at different levels**

Workshops should be held as a leverage to identify concrete research steps and ensure its smooth progress. Five workshops have been organized consecutively in the following forms:

**High-level Workshop on S&T Strategies.** Three workshops on S&T strategies have been organized in October, 2007, December, 2007, and June, 2008, respectively, with the participation of research group heads in eighteen priority areas, chief scholars, and relevant top CAS management members. Information has been exchanged, and consensus been reached to ensure research directions. During the workshops, President Yongxiang Lu pinpointed the significance, necessity and possibility of the roadmap research, and commented on the work of each research groups, thus pushing the research forward.

**Special workshops.** The Editorial Committee invited science policy scholars to the special workshops to discuss the eight basic and strategic systems for China's socio-economic development. Perspectives on China's science-driven modernization to 2050 and characteristics and objectives of the eight systems have been outlined, and twenty-two strategic S&T problems affecting the modernization have been figured out.

**Research group workshops.** Each research group was further divided into different research teams based on different disciplines. Group discussions, team discussions and cross-team discussions were organized for further research, occasionally with the involvement of related scholars in special topic discussions. Research group workshops have been held some 70 times.

**Cross-group workshops.** Cross-group and cross-disciplinary workshops were organized, with the initiation by relative research groups and coordination by Bureau of Planning and Strategies, to coordinate the research in relative disciplines.

**Professional workshops.** These workshops were held to have the suggestions and advices of both domestic and international professionals over the development and strategies in related disciplines.

**(5) Establishment of a peer review mechanism for the roadmap research**

To ensure the quality of research reports and enhance coordination among different disciplines, a workshop on the peer review of strategic research on the S&T roadmap was organized by CAS Bureau of Planning and Strategy, in November, 2008, bringing together of about 30 peer review experts and 50 research group scholars. The review was made in four different categories, namely, resources and environment, strategic high-technology, bio-science & technology, and basic research. Experts listened to the reports of different research groups, commented on the general structure, what's new and existing problems, and presented their suggestions and advices. The outcomes were put in the written forms and returned to the research groups for further revisions.

**(6) Establishment of a sustained mechanism for the roadmap research**

To cope with the rapid change of world science and technology and national demands, a roadmap is, by nature, in need of sustained study, and should be revised once in every 3–5 years. Therefore, a panel of science policy scholars should be formed to keep a constant watch on the priority areas and key S&T problems for the nation's long-term benefits and make further study in this regard. And hopefully, more science policy scholars will be trained out of the research process.

The serial reports by CAS have their contents firmly based on China's reality while keeping the future in view. The work is a crystallization of the scholars' wisdom, written in a careful and scrupulous manner. Herewith, our sincere gratitude goes to all the scholars engaged in the research, consultation and review. It is their joint efforts and hard work that help to enable the serial reports to be published for the public within only one year.

To precisely predict the future is extremely challenging. This strategic research covered a wide range of areas and time, and adopted new research approaches. As such, the serial reports may have its deficiency due to the limit in knowledge and assessment. We, therefore, welcome timely advice and enlightening remarks from a much wider circle of scholars around the world.

The publication of the serial reports is a new start instead of the end of the strategic research. With this, we will further our research in this regard, duly release the research results, and have the roadmap revised every five years, in an effort to provide consultations to the state decision-makers in science, and give suggestions to science policy departments, research institutions, enterprises, and universities for their S&T policy-making. Raising the public awareness of science and technology is of great significance for China's modernization.

Writing Group of the General Report

February, 2009

# Preface

The research project *Information Science & Technology in China: A Roadmap to 2050* formally started in November 2007 with a research group composed of 13 experts. On January 11, 2008, the research group held the first workshop at the Institute of Computing Technology of the Chinese Academy of Sciences, when the general framework of the information science & technology (IS&T) roadmap was completed. Between February and October 2008, four workshops were held by the research group. In September 2008, the first draft of the IS&T roadmap was composed. Based on recommendations and comments from an advisory committee headed by academician Hu Qiheng and the overall requirement of the Chinese Academy of Sciences, the draft was revised substantially. In August 2009, the formal manuscript of the research report was submitted for printing.

All the experts in the research group share a consensus that, due to the rapid paces of progress in IS&T, with many unanticipated potential breakthroughs, it is difficult to make an accurate forecast of the development of IS&T for the next 40 years. Our strategic research on the roadmap is based on our understanding of the national strategic requirements and the basic laws of S&T development. We analyzed the trends and constraints of IS&T development in the decades to come; and then, offered our judgments and suggestions on important strategic priorities that China should select. We do not think that the development of science and technology is a spontaneous process completely independent from human efforts. Rather, the future of IS&T is heavily affected by research and development activities guided with objectives of human beings.

Based on our understanding of national strategic requirements and trends of information science and technology, we select the following areas as our focus of the roadmap study: network science and future network technology; microelectronics, optoelectronics, quantum IT; supercomputing, software and information storage; knowledge based technology and future information service based on knowledge processing; low-cost information systems and upgrade of traditional industry by IT; interdisciplinary sciences of intelligence and cognitive science, bioinformatics and social informatics; fundamental theory of information science; and information security.

As a whole, our team was not very familiar with the method on roadmap research. Generating this report was also a learning process for us. We look forward to receiving comments and feedbacks from experts both inside and outside this arena, and hope that the research results presented in this report can be useful as a reference, for the long-term planning and research direction selection of S&T in China.

Strategic Study Group on IS&T of the
Chinese Academy of Sciences

October, 2010

# Contents

# Abstract

In the first half of the 21st century, important changes and paradigm shifts are emerging in information science and technology (IS&T):

(1) Information technology (IT) is entering a stage of mass adoption, and it is possible that we will see in the 21st century the "Cambrian explosion" of information technology applications. Computing for the masses will become the main theme in the next several decades.

(2) Saving energy and reducing pollution have become an important requirement in the development of IS&T. More attention will be paid to sustainable development and social harmony.

(3) In the next 10 to 15 years, Moore's Law that has been valid for the past 50 years will face unprecedented challenges.

(4) The existing arena of IT is changing from a man-machine symbiosis to a ternary universe of the cyberspace, the human society, and the physical world. Traditional IS&T is to be adapted to suit developing application systems in such a ternary universe.

(5) IS&T is penetrating various application areas and intersecting with biology, nano science, and cognitive science, to form new disciplines. Computing is becoming the fabric tying various scientific and technological disciplines together.

In our roadmap, from 2010 to 2050, the overall goals of IS&T development in China can be described as the following:

Playing an active and substantial role in the transformative change in information science and technology; Enhancing our capabilities of innovation and sustainable development to enable China to become a universal information society (U-society), in which most part of the population will be users of information systems, information will be the most important resource for the economy and the society, and the development level of information systems and their application in China will be close to the developed countries.

We suggest six major tasks and objectives to be focused on till year 2050:

(1) Constructing a ubiquitous, well-content information network.

(2) Realizing revolutionary upgrades of information devices and systems.

(3) Developing a data and knowledge service industry.

(4) Upgrading traditional industries by IT and realizing low-cost informatization.

(5) Developing new information science and interdisciplinary sciences based on computation.

(6) Constructing a national and social information security system.

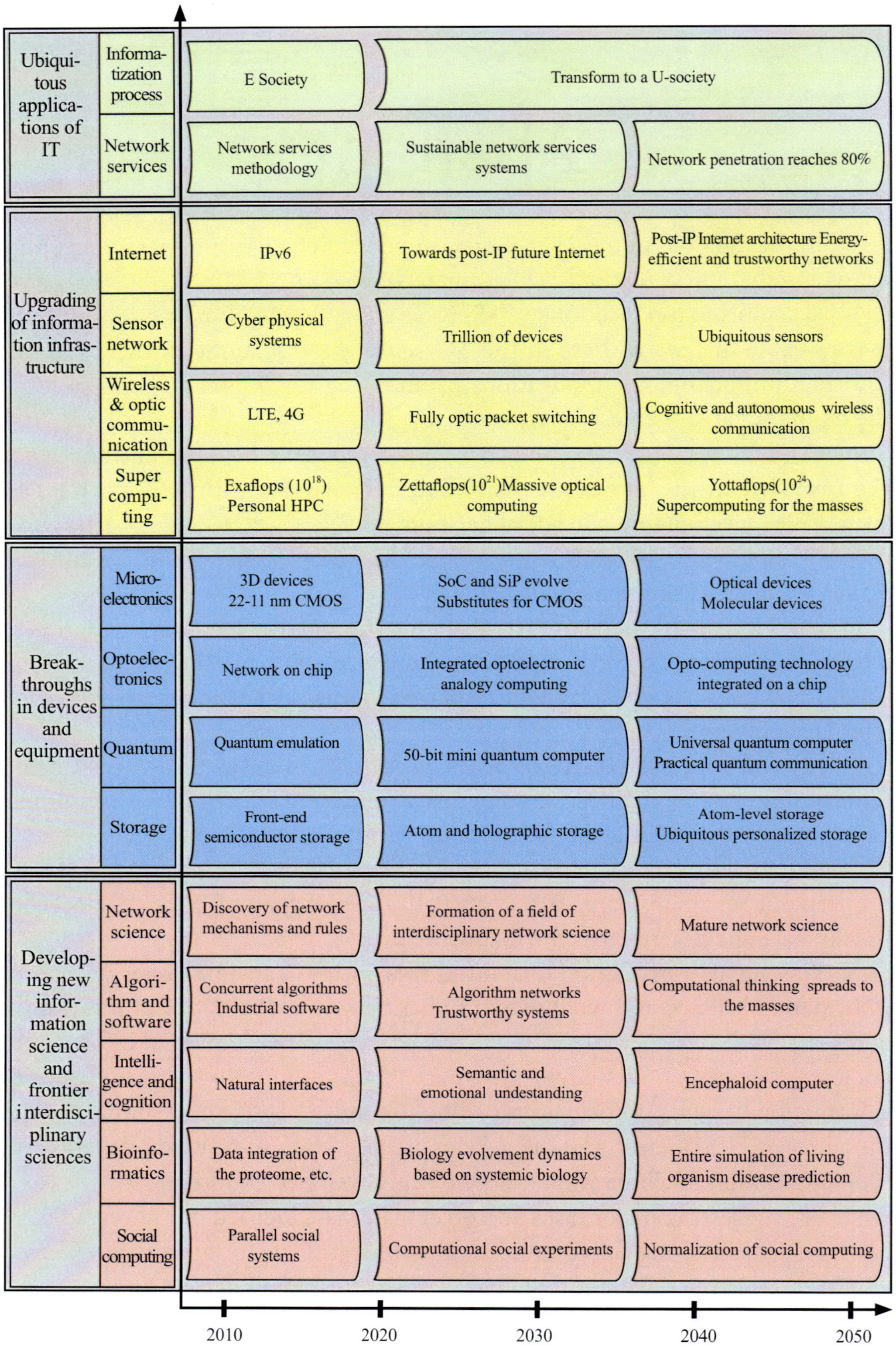


Figure 0-1　A Roadmap to 2050, IS&T development in China

All these tasks can be combined into one overall goal: to establish a

Universal, User-oriented, and Ubiquitous Information Network System, or U-INS system. This system is to meet the strategic demand for China to enter an information age. The implementation of the U-INS system covers China's scientific and technological priorities in the first half of the 21st century. We have to carry out innovative research in various areas, including transformative devices, new generation network systems, personalized network services and network applications, network security, network science and new information science. Figure 0-1 shows the general roadmap to realize these objectives.

The following are the main findings and recommendations derived from our roadmap research:

(1) The second half of the 20th century was marked by inventions and innovations in information technology. However, during this period there were few fundamental breakthroughs in information science. Due to this lag in information science, the incremental advances in key technologies, such as IC and network will no longer be able to continue beyond 2020. We foresee that 2010–2030 will be a period of breakthroughs in information science. The first half of the 21st century will see an information science revolution, most likely in the areas such as network science, intelligence science and computational thinking. The science breakthroughs might trigger a wave of information technology revolution in the second half of the 21st century.

(2) The number one task for the next 10 to 40 years is to build an information network that enables people to conveniently access information and knowledge, to more effectively cooperate, and to attain higher quality of life. In the next 10 years, network technology will evolve in different dimensions including broadband networks, mobile networks and Integration of Three Network (Internet, telecommunication network and television broadcast network), towards the IPv6-based Internet. After 2020, countries in the world will gradually reach a consensus on constructing a Post-IP network system architecture. Broadband wireless communication is a cornerstone of the future network systems. Ubiquitous sensor networks will combine with the space, ground and access network systems, to enable human-human, machine-machine and human-machine communication anytime and anywhere.

(3) We can not take information science and technology only as a high-tech tool. Instead, we have to deeply understand the ternary universe of human-computer-thing (or human society, cyberspace and physical world as mentioned above). In the process of developing information science and technology, we need to solve problems in the following six fundamental aspects: hardware development, large scale parallel programming of the ternary universe, efficient utilization of massive data, construction of low-cost information network, establishing trustworthiness of information system, and construction of a cyber-infrastructure free from monopoly.

(4) Traditional information devices and equipments, which have problems with ever-growing complexity, cost, and energy consumption, are calling for disruptive technologies. There exists no definitive roadmap for

the future similar to that of the CMOS in the past 30 years. Quantum, self-spin and nanotechnology all exhibit uncertainty and diversity. It will probably take 15-20 years to ascertain the mainstream of device technology. Transistors based on graphene are likely to be a propeller for extending Moore's Law, and may be a promising research direction to help us move beyond Silicon-based CMOS. The combination of electronic, optoelectronic, and optical computing technologies will most likely lead to new chip technologies integrating memory, communication and information processing capabilities. Optical interconnection and large scale optical computing would also be able to be realized on this kind of new chips.

(5) By 2050, super servers, with much higher performance and capacity, are required to support various personalized application workloads. To meet these requirements, a series of technical bottlenecks such as power consumption, massive parallelism, reliability and cost have to be broken. The speed of supercomputers will be raised by $10^8$–$10^9$ times, to $10^{24}$ flops within 40 years. Major breakthroughs are called for in the stage from Exaflops ($10^{18}$ flops) to Zettaflops ($10^{21}$ flops). Another important goal is to enable the service and software sectors to achieve growth in a way similar to Moore's Law, that is, to decrease the cost of the same software and service by 50% every two years.

(6) When looking from a historical perspective, the diffusion rate of computers is about the same as that of electrical power. Computing for the masses should not sacrifice effectiveness and value, and value should grow proportionally to penetration. Mass adoption of computing implies not only the increase of low-cost users, but also the increase of high-value ones. Our study shows that computing for the masses have to follow a value-augmenting path, which not only is an effective low-cost path but also facilitate the IT industry growth. A universal compute account is to be provided to each of the 1.2 billion Chinese users, so that any user would be able to access and use his or her personalized information environment conveniently and effectively, with various devices, anywhere and anytime.

(7) Human-computer interaction is a main focus in computer science and engineering research. In the coming decades, multimodal human-computer interactions will become main-stream in desktops, laptops and palm systems. 3D user interaction, tangible interaction, personalized emotional interaction, and brain-computer interaction will be popularized. Natural language understanding and image semantic understanding are important and difficult issues to be tackled in the long term.

(8) A bottleneck of both the Internet and information service is the semantics understanding capability of computers. Developing semantic Web technology is an important way to realize a truly universal Internet. We must explore and utilize the characteristics of Chinese culture, to develop techniques and network platforms that support semantics, content, and culture. We should expand the portion of Chinese contents in the Web to more than 10%, thus provide a necessary base for developing a data and knowledge industry with

Chinese characteristics.

(9) Quantum information can provide new principles and methods for developing information science and technology, and will probably become one of the new post-Moore's Law technologies. The difficulty of quantum computing implementation is not in theory, but in physical realization. Quantum computers are likely built on the basis of solid physics and quantum optics. Quantum cryptography technology has been utilized in the engineering research and close to application. It is anticipated that quantum cipher-key distribution in metropolitan fiber networks within 70 km will be realized by 2020, in practical global communication networks by 2050.

(10) In network computing environment, which is featured with distribution, interaction, and parallelism, new computing models and new theories for the design and analysis of algorithms are required. A big challenge in the coming decades is to establish rigorous mathematical models and a solid theoretical foundation for concurrent computing. Algorithmic research will shift from single algorithm design to interaction and collaboration of multiple algorithms. Computing software is becoming larger and more complex, causing a decrease in reliability and security. Establishing software foundation for dependable computing has become a scientific problem that must be solved in the next several decades.

(11) One of the most challenging fundamental scientific themes of our time is exploring the nature of human intelligence, understanding the brain and its cognitive functions. Breakthroughs in cognition-based intelligent information processing have the potential to lead to great progresses in IS&T. Developing new intelligence science and technology is an important goal for the coming 50 years. Brain reverse engineering and brain-computer interfaces are worthy research directions.

(12) Modeling the development process of cells from a computational perspective not only benefits the understanding of such fundamental issues as how genes and proteins interact to control the metabolism of cells and the renovation of DNA in an organism, but also benefits the study of communication protocol design, concurrent computing models and mechanisms design. Research on information transformation processes at molecule and DNA levels may lead to the emerging of new computing systems different from silicon based ones.

(13) Social computing has become a frontier research topic in computing following scientific computing and bio-computing. Research on social computing based on cognitive science, intelligence science, and complex system science is becoming a priority for national security and the construction of a harmonious society.

# 1 Trends of IS&T in the First Half of the 21st Century

## 1.1 Overall Trends of IS&T

In the coming decades, IS&T will continue advancing at high speed, further extend its influence and penetration, change our economy and life styles, and have a deep effect on learning, entertainment, governance, enterprise operation, cultural diffusion, and so on. Computers and communication networks will achieve new advances in terms of speed, capacity, bandwidth, convenience, reliability and security. Disruptive changes in information technology will come after fundamental breakthroughs in information science.

### 1.1.1 IS&T development will place more importance on mass adoption, sustainable development, social harmony and the openness of the industry

(1) While paying attention to core technology breakthroughs, we should pay much more attention to applications value, penetration, and mass adoption of IT, particularly to important issues like narrowing the digital divide, benefiting the masses, lowering cost of information use, enhancing ease-of-use, stability and security of information products and services. If we can accomplish the above, most likely a *Cambrian explosion* of information technology applications will occur in the 21st century as predicted by some experts.

(2) While concerning about roles of IT in increasing competitiveness and economic benefits, accompanying the knowledge economy age, we need to attach more importance to the impacts of IT to ecology and environment. We also need to explore ways of sharing limited natural resources as well as unlimited knowledge resources with the help of IT, when pursuing sustainable development.

(3) While attaching importance to the combination of information science and technology, we need to attach more importance to the combination of information science and social science, of IT and arts and humanity; pay attention to the ethics and morality issues in IT research, the positive and

negative effects of IT on society.

(4) While developing economy of scale of IT, we need to attach more importance to its diversity and openness. In the next half a century, for the healthy development of IT, we need to form an open, collaborating and resources and benefit sharing industrial ecosystem, breaking through the limitations of the current markets and intellectual property restrictions to create new and larger markets, new mainstream technology, and new forms of intellectual property rights.

### 1.1.2 IS&T will integrate with various application domains and become a bond to link different scientific disciplines

(1) IS&T will be gradually integrated with other technological domains. Many future advance of IS&T will emerge from interdisciplinary technological innovations, instead of vertical upgrades only. Thus we should not only focus on the so-called "key" or "mainstream" IT technology.

(2) While striving to make breakthrough in the core and critical technologies in the IS&T, we need to attach more importance to the exploitation of new sciences, especially to the cross-discipline research of IS&T and technical areas like nanotechnology, life and cognitive science, in order to realize convergence development. New sciences, such as bioinformatics, social informatics and social computing, and nano-intelligent science, will be developed based on IS&T, especially through computer simulations.

(3) In the next half a century, breakthroughs in the theory and method of intelligent information processing based on cognitive mechanisms have the potential to lead to disruptive transformations in IS&T. The convergence of brain science, cognitive science and artificial intelligence will solve grand theoretical problems in cognitive science and bring IS&T into a new age with the characteristics of brain simulation (including brain reverse engineering).

### 1.1.3 New changes in IS&T

(1) Moore's law reflects the development of IT in the $20^{th}$ Century. One of the important goals of IS&T in the $21^{st}$ century is to make the information systems, including application software, solutions, and services, generate a scale effect, thus to enable the software and service industries to develop in a way similar to Moore's law. For example, the cost of software development with the same function and capability will decrease by 50% every two years, and so will the cost for the same service. In the coming decades, if we can enable the software and service industry to have "Moore's law"-like development, a revolution will undoubtedly be initiated.

(2) Within ten to fifteen years, Moore's law for CMOS will still function. When the feature dimension of IC technology gets to be less than 10 nm, new chip technology such as carbon nano devices and molecular and quantum devices, could become mainstream. The combination of electronic computing, photonics and optical computing technologies most likely will create a new generation chip

technology that fuses computation, memory and communication, and might lead to the realization of optical computing on chips, hence to the development of the technology to support "Internet on chip" and "labs on chip".

(3) Following the system theory, information theory and cybernetics created in the $20^{th}$ century, a new network theory (Net theory) will be created in the $21^{st}$ century. It will regard the whole Net as a complex giant system, and its development will have a profound effect on the popularization of the Net. It will also bring the algorithmic research from focusing on single algorithms to interaction of multiple algorithms, and will establish a new theoretical foundation for the development of distributed systems.

(4) With the increasing use of a variety of embedded devices and sensors in the information systems, the ratio of the edge devices to servers will increase by several orders of magnitudes. It will be a new challenge to store, search, check, gather and analyze the information sent by the embedded devices and sensors in the $21^{st}$ century.

(5) In the past several decades, the development in information science and technology presents a characteristic that competing technologies developed in parallel, and dominated the market alternately in different periods. For example, development of computing platforms showed cycles of consolidation-decentralization. Another instance is the alternative development of general purpose integrated circuit (IC) and application specific IC over a period of ten years. Every mode change is not a simple negation of the former one but rather a spiral up. Research on the rules of the macroscopic development will decrease the one-sidedness in the research on the roadmap and in decision-making.

(6) Since the 1970s, digital technique has been a primary technology in the information field. But as we put the most emphasis on digital technology the analog one should not be ignored. When signals are changed from analog to digital via discretization, combinatorial explosion cannot be avoided in many applications. Digital computers can cope with just a small part of the problems in the real world accurately. The solvable problems have to be of low complexity and able to be formalized. In the next several decades, analogue computing might become a topic of research again. We should explore new ways for analog computing, and find new methodology for digital-analogue hybrid processing, as we emphasize digital technology.

## 1.2 The Next 20 to 30 Years will be a Period of Transformation and Breakthrough for IS&T

Information science is still young. In the past 30 years, information technology developed faster than information science. Many important fundamental problems of information science have not yet been solved. In the past 20 years, with the development of IC and Internet technology, some challenging problems related to information science have been brought up. Some important issues are: How to deal with the complexity of the complex information systems (multi-core chips, large-scale software systems, networked systems)? Currently energy costs of information systems are far greater than the theoretical lowest bound. How to get revolutionary devices to decrease energy costs by several orders of magnitudes? Reliability and security have been main obstacles for the development of IT systems all the time. How to develop theories for security and reliability that can effectively direct the construction of the information system? The exploration of these challenging problems coupled with the wide application of IT will further promote information science.

The second half of the 20th century was characterized by the invention and innovation of information technology. It is predicted that the first half of the 21st century will see a revolution in information science, characterized by network science, high performance computing and simulation, intelligent science and computational thinking. The breakthroughs may result in a new IT revolution in the second half of the 21st century, as shown in Figure 1-1.

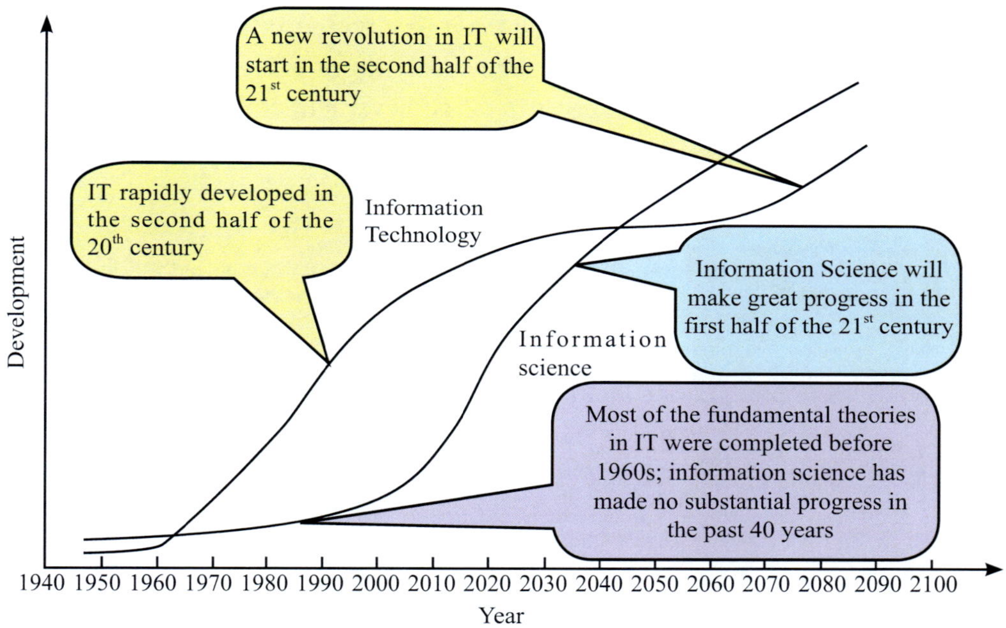


Figure 1-1 The predicted long-term development of IS&T

CMOS technology that is the basis for most computer systems has faced a fundamental challenge: the future of this technology after 2020 is not clear. We should not only continue to develop microelectronics, optoelectronics and photonics devices based on new technologies such as nano technology and superconducting, but also investigate how to effectively use new technologies such as quantum and biology on the layer of computational models and computer architectures. These technologies have different properties (as shown in Table 1-1), which have a direct effect on both architecture and parameter choices. Future computers and information systems can adopt a mixture of different technologies, including new semiconductor circuit technology (e.g. 3D circuit technology), optical technology (e.g. optical interconnect), nanotechnology (e.g. memory and displays) and quantum computing technology, etc. The next 20–30 years will be a transformational period for information devices and systems.

**Table 1-1 A comparison of new technologies (Gate, 2016) (ITRS ERD Section)**

| Technology | Speed | Size | Energy cost |
|---|---|---|---|
| CMOS | 30 ps–1 ms | 8 nm–5 nm | 4 aJ |
| RSFQ | 1 ps–50 ps | 300 nm–1mm | 2 aJ |
| Molecular | 10 ns–1 ms | 1 nm–5 nm | 10 zJ |
| Plastic | 100 ms–1 ms | 100 mm–1 mm | 4 aJ |
| Optical | 100 as–1 ps | 200 nm–2 mm | 1 pJ |
| NEMS | 100 ns–1 ms | 10–100 nm | 1 zJ |
| Biological | 100fs–100 ms | 6–50 mm | 0.3 yJ |
| Quantum | 100 as–1 fs | 10–100 nm | 1 zJ |

Notes: For the multipliers of m-u-n-p-f-a-z-y, m means $10^{-3}$, u means $10^{-6}$, n means $10^{-9}$ and so on till y means $10^{-24}$, proceedings in increments of three orders of magnitudes ($10^{3}$).

Research on information systems can be divided into four layers: ideas, principles, theories and tools. On each layer, there exist many challenging scientific problems. Figure 1-2 shows great achievements in the past as wells as some research directions that have the potential to make great breakthroughs in the coming decades.

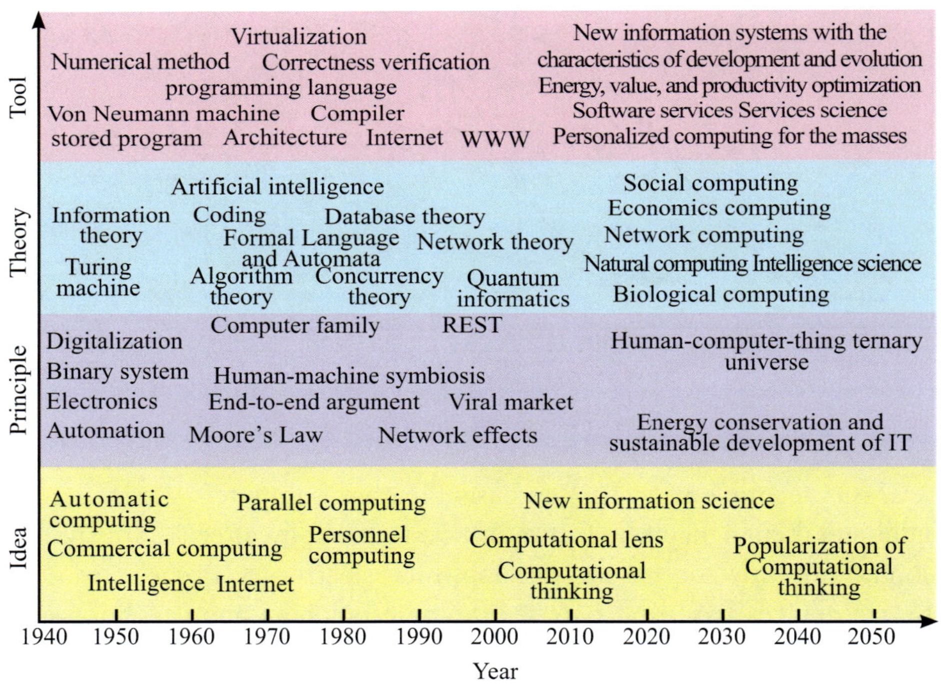


Figure 1-2 Key areas for comprehensive research on information up to 2050

## 1.3 IT is Entering the Mass Adoption Phase

The first computer appeared more than half a century ago. IT has passed three phases, namely expert use, early adoption, and public recognition, and is entering the mass adoption phase from the beginning of the 21st century. New technology will continue to emerge in the IT area, and some of them might be disruptive. According to the Long Wave Theory of Nikolai Dimitrievich Kondratiev, a Russian economist, every long economic wave will last about 60 years. The present wave based on the technology of computers and the Internet has passed its halfway. Therefore, the main stream of the development of IT in the next 30-40 years will be mass adoption and application of the technology, particularly making IT benefit the masses.

One popular opinion states that for the information technology products, such as TVs, computers and the Internet, the time from their invention to mass adoption, has clearly shortened from several decades to several years compared with traditional technology products. This widely accepted opinion maybe ignored the difficulty of the mass adoption of IT. Actually, from the viewpoint of long-run, we may find that the speed of popularization of the computer (from 1971 to 2001) is almost the same as that of electric power, as shown in Figure 1-3. Thus, a fundamental challenge for all the people in information field is to promote ease-of-use. Compared with the plug-and-play ability of electric power products, IT products have a long way to go to gain the same ease-of-use. A breakthrough in information science will probably come from the progress of intelligence technology. The more intelligent the machine is, the more convenience the user will gain.

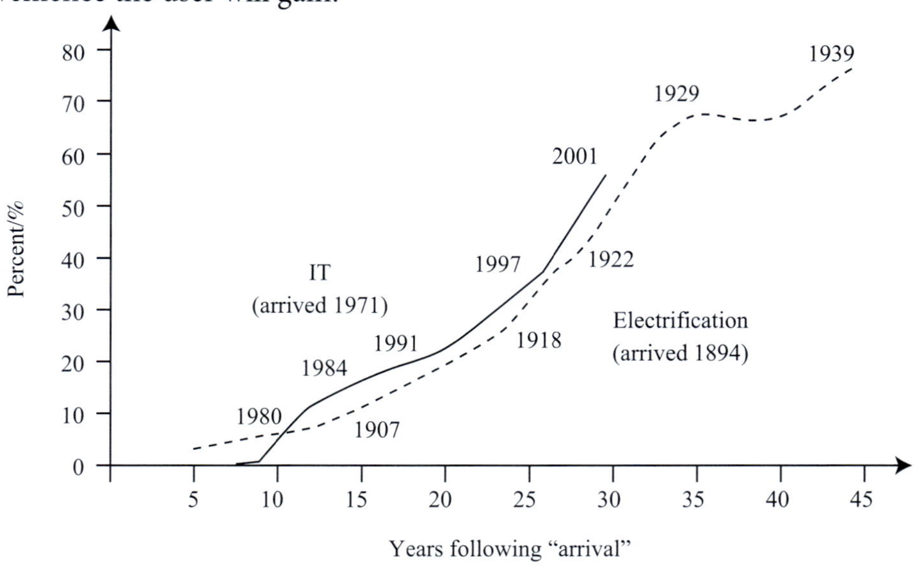


Figure 1-3 The popularization speed of the IT and electric power are almost the same
Source: Jovanovic and Rousseau, General Purpose Technology, National Bureau of Economic Research, 2005. http://www.nber.org/papers/w11093

What are the main driving forces of IT popularization? Besides the enterprises, which play a dominant role in innovation, the individual consumer is also an important one. The popularization of computers, the development of net computing and the spread of open source software will make low-cost platforms available to enable users who have development capability to form communities by themselves. The end user would not only be a consumer but an inventor in IT and a source of information creation. Community will be another force for IT innovation, especially in software domain. Utilizing these three forces will significantly improve the effectiveness of innovation in the information field, and expand application and market for IT products, especially software and network services.

## 1.4 IS&T will become the Bond of Various Sciences for Discipline Crossing and Convergence

Since the founding of experimental science in Europe in the 1700s, the fundamental methodology for scientific research has been reductionism, which divides the real world into smaller (also simpler) parts or profiles and then studies them. The studies of these parts and profiles shape the difference of various scientific branches. Information science is just one of them. The division of sciences is so fine and separated that it obscures the global and integrated understanding of things. The renowned German physicist Plank believes that *science is internally a unit, and divisions of science are not dependent on themselves but on the insight location of the human being. Actually, there is a permanent link between Physics, Chemistry, Biology, Anthropology and Sociology, and it is a chain that can not be broken in any place.*

To cope with more and more complex problems, many researchers began to search for the chain that was broken, using the integration method. When giving a keynote speech at the Supercomputing 2002 international conference, Dr. Rita Colwell, then director of the US National Science Foundation, illuminated a main trend for scientific research in 21$^{st}$ century: the shift from reductionism to integration.

In the past several years, NIBC, the convergence of nano, biology, information and cognitive sciences becames a hotspot in science and technology. The convergence of any two, three or four technologies in NIBC will generate important impact on the development of science and society. For example:

(1) IT will provide new avenues for research of material science, life sciences, astronomy, geosciences, energy resources, ecology and environmental sciences and technologies. New interdisciplinary sciences and technology frontiers taking computation as the core will emerge.

(2) Computer science is important to life sciences just as mathematics to

physics. The thinking modes of computer science will permeate into the overall processes of molecular biology research. Computational biology will become one of the main branches of biology.

(3) Quantum physics, nanotechnology, and biotechnology will provide IT with new materials and devices for information storage, transmission, processing, and display.

(4) Progress in brain science and cognitive science will likely lead to breakthroughs for information expression and processing. Intelligence technology based on the achievements of brain and cognitive science will cause a new revolution in IT.

Currently the computers with the highest performance in the world are used in the simulation research of resources and environmental sciences, nano and life sciences and new energy resources. IT, especially simulation based on high performance computing, has been a necessary bond for interdisciplinary research. There was a popularized view that *computer simulation is the third kind of tools to conduct research other than experiments and theoretical analysis*, but now it seems not only so. In the development trend of science in the 21$^{st}$ century, the computer has changed from a tool serving traditional research scientists to the very fabric of sciences. That leads to a new research paradigm of *computation + traditional sciences = new sciences*. In 2008, Turing Award winner Richard Karp said to the graduate students of Chinese Academy of Sciences that *the algorithmic worldview is changing the sciences: mathematical science, natural sciences, life science, and social sciences. Computer science is placing itself at the center of scientific discourse and exchange of ideas.*

With the development of IT, more and more new sciences with the styles of Compu + X or X-info have emerged, such as computational physics, computational chemistry, computational biology, computational sociology, bioinformatics, and nano-informatics, etc. IS&T play a critical role to bind various sciences together, and without it interdisciplinary sciences can not develop well. The ability for processing and recognizing the patterns and information hidden in massive data will help us unlock the doors to understand mysterious processes in various fields from biology to society. Many of life phenomena including propagation, growth and self-restoration and so on are processes that can be decoded and emulated as computing. In the next 40 years, the popularization of high-performance computing will lead to the popularization of computer emulations. The emulations can be of great help in the process of decision making not only in complex scientific or social issues, but also in daily life.

Ancient Egyptians measured the ground after each flooding of the Nile, which led to the development of geometry and algebra. The massive routine work of computer programming will probably lead to the appearing of new mathematics. As the invention of the telescope aided astronomy and the microscope promoted the field of medicine, the invention of digital computers, especially the rapid development of microprocessor and network technology in the past 20 years, have made massively parallel computing and network computing possible. This will cause a revolution in science. "New sciences" based on parallel computation will appear in the 21st century.

The convergence of various kinds of sciences and computers is based on mathematical models that can be programmed and computed. In the sense of methodology, mathematical modeling currently is at the "level of hackers". The binding of experimenting and modeling demands a kind of methodology to organize the data and computation resources, and a mature framework to manage the models and complex dynamic relationships. New data models, system architectures and sophisticated control mechanisms are needed in the interdisciplinary research to support the workflows of all kinds of sciences, and to integrate the flows into an intertwined scientific process.

Besides market demand, another impetus for the development of IS&T is driven by the intrinsic progress of science and technology. The principal technology for the first industry revolution was steam power. Those for the second were internal combustion engine and electrification, and for the third, which is still going on, are electronics, information and network. Future technology progress will not be in only one or two areas, but rather groups of different technologies such as IT, biological technology, nanotechnology, cognitive science and intelligence technology, new materials and advanced manufacturing technology, aviation and aerospace technology, new energy and environment protection technology. These technologies will enter into a new age of convergence and co-evolution. New inventions will focus on interdisciplinary fields. Convergence between the sciences will become more frequent.

## 1.5 Inspiration from the IS&T Strategies of Developed Countries

The United States, European Union and Japan are all placing high emphasis on the strategic research in the information field. Though no roadmaps of IS&T to 2050 of these countries have been published, we can still gain some insights from their published development strategies of IS&T.

In the information field, the support of the US government can be seen from the Federal Networking and Information Technology R&D Program (NITRD). In 2008, more than 3.3 billion US dollars were invested into this program. The program has seven topics including cyber security and information assurance; human computer interaction and information management; high confidence software and systems; high end computing; large scale networking; software design and productivity; social, economic, and IT workforce development.

From the topics mentioned above, we can understand the US government's focus in the information field. In the past several years, research has emphasized forward-looking and fundamental works. The US National Science Foundation began research on computational thinking in 2007, and the revolutionary, transformative and paradigm-changing was further emphasized in the guideline of the program in 2008.

The seventh framework programme (FP7, 2007–2013) of the development of science and technology published by the European Union showed the major trends and key points of European countries. Europe's main objectives with FP7 in information and communication technology (ICT) are *improving the competitiveness of European industry and enabling Europe to master and shape future developments in ICT so that the demands of its society and economy are met.* The research activities focus on the improvement of the software and hardware of different ICT and the combination and application of different technologies. Funds for ICT research reached 9.11 billion Euros, ranking number one among the funds in FP7. The main research activities can be separated into several categories:

(1) ICT Technology Pillars: nano-electronics, photonics and integrated micro/nano-systems; ubiquitous and unlimited capacity communication networks; embedded systems, computing and control; software, grids, security and dependability; knowledge, cognitive and learning systems; simulation, visualization, interaction and mixed realities; new perspectives in ICT drawing on other science and technology disciplines, including insights from mathematics and physics, biotechnologies, material and life-sciences.

(2) Integration of Technologies: personal environments; home environments; robotic systems; intelligent infrastructures.

(3) Applications Research: ICT meeting societal challenges; technology for health. To improve inclusion and equal participation and prevent digital divides; technology for mobility; technology for risk management and sustainable development, disasters prevention; Technology for governments at all levels.

(4) ICT for Content, Creativity and Personal Development: new media paradigms and new forms of content; technology-enhanced learning; ICT-based systems to support accessibility and use over time of digital cultural and scientific resources and assets.

(5) ICT supporting businesses and industry: new forms of dynamic networked cooperative business processes; manufacturing; ICT for trust and confidence.

From the key areas of FP7, we can see that European countries place high importance on the integration and application of technologies. An interesting phenomenon is that communication, software and cognition technologies that we regard as the core technologies are classified into the task of *an integration of nano-electronics, nano-optics and a combination of micro/nano systems.* This shows that European countries attach more importance to nano technology and have high hopes for it. A document called *Moving ICT frontiers—a strategy for research on future and emerging technologies in Europe* was published in April 2009 by the European Union, and put forward a goal *to improve the competitive power and ecological environment for innovation by increasing investment in strategic research with primary risks.*

The Japanese government has been promoting the development of ICT all along. Having fulfilled the e-Japan plan ahead of schedule, Japanese government initiates the u-Japan plan with high priority. In the plan, the concept of "Ubiquitous" is denoted by a "U" along with three "u"s (universal, user-oriented and unique). The "U" focuses on infrastructure construction directed by technology development to connect "every one and every thing anytime, anywhere, by anything and anyone" The three "u"s focus on the expectation of the future scenario: universal means that anyone including elderly and disabled will be able to use ICT with ease; user-oriented means commerce involving product sales and services will shift from being provider-oriented to user-oriented, based on users' viewpoints; unique means ICT will also transform society from one of uniformity and standardization to one that is creative and vigorous, and which strives to achieve more creative business approaches and services, as well as a new social system and values. The strategy of u-Japan focuses on service while e-Japan focused on infrastructure and production. One core of the Japan strategy, which presents an important trend in the information industry in the future, is the convergence of the information industry, information services, data content industry and even the solution of related social problems, and it is also one of the cores of u-strategy.

South Korea has undergone a similar developing process. With a leading position in mobile communications, home electronic appliances and data content services, it has the highest coverage of the broadband service in the world. To cope with the new trend of global information, South Korea created their own u-Korea strategy embodied in the IT839 strategy drafted by the Ministry of Information and Communication. The report *Humanism in the Digital World IT839 Strategy* states that *a ubiquitous network society is a kind of technological society armed by intelligence networks, most advanced computing technology and other advanced digital infrastructure.* The u-Korea strategy emphasizes that the development of IT and information services should not only meet with the demand of industry and the economy but also make a revolutionary change in people's life. The shift from *e* to *u* will help to realize many dreams that could not be realized in e-time.

Developed countries place high importance on the research of foresight and interdisciplinary sciences. In Aug. 2007, a report on *Leadership under Challenge: IT R&D in a Competitive World* was made by the President Council of Advisors on Science and Technology of the United States of America. This report pointed out that short-term and incremental research was overemphasized in the IT projects supported by the US government, and advised that the support to the projects of long-term, with more inherent risks and across disciplines should be enhanced, and universities should reconsider their academic structure and incentive mechanisms and promote cooperation of larger scale across different disciplines. It also requested collaborative research projects across different branches of government to maximize the efficiency of national R&D investments to keep the competitive edge of the US.

Actually, the US government has already enhanced the support for the research on foresight and interdisciplinary research. GENI (Global Environment for Network Investigation) is one of such cases in the research of network. Its mission is to open the way for transformative research at the frontiers of network science and engineering; and inspire and accelerate the potential for groundbreaking innovations of significant socio-economic impact. It is planned to build a network environment beyond the capabilities of and not necessarily compatible to the current Internet so that researchers can conduct disruptive experiments at-scale. The GENI plan is functioning as a platform and test bed for the research and experiments in the Internet more or less in the way like a particle accelerator functioning in the physics research. In addition to GENI, a new program called FIND (Future Internet Design) has also been initiated by the NSF. It invites the research community to consider what the requirements should be for a global network of 15 years from now, and how we could build such a network if we are not constrained by the current Internet. The plan tried to find what the future network will be, how better in terms of security, availability, manageability and impact to the economy and social development it will be than the existing one, and how to design and construct it. The new network architecture will be developed by FIND and run on the GENI experimental network.

Though GENI did not get the expected budget of several hundred of million USD from Capitol Hill, and its actual progress was slower than expected, it provided some inspiration on how to create a roadmap for S&T and to free our mind from the cage of existing technologies when making choices on important technologies that may have profound effects on human society.

The US Department of Energy plan INCITE (*Innovative and Novel Computational Impact on Theory and Experiment Program*) gives us some inspiration on how to improve our support to interdisciplinary research. It offers computer hours of the world's most powerful supercomputer operated by DOE to teams all over the US and the world on a competitive basis.

ITRS is a template for the roadmap of semiconductor technological development. Since the appearance of IC, there has been anticipation that the size of micro-electronic devices would be proportionally shrinking and so would the cost per unit capacity or performance. The market would likely expand accordingly as well. In the competitive IC market, there is one question that must be answered for all practitioners:=which technologies can make the semiconductor industry keep the development at a Moore's law pace. To answer this question, the US Semiconductor Industry Association (SIA) compiled the National Technology Roadmap for Semiconductors (NTRS). Three versions of the document were compiled consequently in 1992, 1994 and 1997. In 1998, SIA invited researchers from Europe and Asia to update the roadmap, thus creating the first version of The International Technology Roadmap for

Semiconductors in 1999 (ITRS). From that point forward, ITRS was compiled in odd years and updated in even years. The overall purpose of ITRS is to offer predictions on research demands for the next 15 years that are identified by the industrial community. Therefore, the conclusions of ITRS are important for all stakeholders like industry, academic community and governments, critical for decision-making at all levels, and of help to guide R&D efforts to the right direction where breakthroughs will most possibly be made.

# 2 The Strategic Requirements for IS&T in China

In order to study the roadmap, we need to understand the current status of China and the strategic requirements of IS&T from the economy and social development better. The demand of the country and its people is the fundamental motivation for the development of IS&T.

## 2.1 The Status of IS&T Development in China

In the past 50 years, especially since the reform and opening up, IT in China has made great progress, which is reflected by the current scale of the industry and IT users. According to the Ministry of Industry and Information Technology of People's Republic of China (MIIT), in 2008 the sales revenue of the information industry reached 6300 billion *yuan*, an increase of 1490 billion *yuan* from the previous year, or 14.6 percent growth. The portion of value added from the information industry in GDP reached five percent. The scale of the information industry occupies a leading position in the economy, becoming the No.1 pillar industry in China.

In 2008, the production of mobile phones in China reached 560 million sets, an increase of 2 percent from the previous year; the color TVs reached 90.33 million, a 6.5 percent increase; PCs reached 137 million, an increase of 13.2 percent; digital cameras reached 81.88 million, an increase of 9.3 percent; IC chips reached 41.7 billion, an increase of 1.3 percent. Telephone subscribers reached 982 million, and Internet users reached 298 million. The scale of Internet in terms of users is number one in the world. The mobile phone penetration reached 48.5 percent, while color TVs and PCs in cities reached 150 percent and 60 percent, respectively.

In 2007, R&D investment in information industry in China reached more than 100 billion *yuan*. The ratio of R&D to sales of the top 100 IT companies as a whole was about 4 percent. R&D expenditure of Huawei Co.

reached more than five billion *yuan*. By the end of 2006, there were 1.51 million pending and awarded patents in China. Among them 460 thousand (i.e. more than 30 percent) were from information industry. In 2008, Huawei Co. applied for 1737 PCT international patents, becoming the number one applicant in the world of the year. In addition, there were 172,000 Chinese academic publications, the second highest in the world for the year and 8.4 percent of the world's total. In 2007, there were 1.74 million undergraduate students in software and 136,000 graduate students, thus the number of IT professionals educated in China is in a leading position in the world. All these show that the scale of both R&D activities and the industry of IT in China are considerable.

In spite of the fast growing, China's IT industry is not competitive enough in the world market. In the past several decades, there were dozens of "disruptive" inventions in IT, including IC, RISC architecture, the Internet, Web and its browser, UNIX and Linux operating systems, graphical interface, mouse, etc., none of which was invented in China. In spite of the fast growing publications, few are presented in top conferences or top journals. The number of patents applied for in China dramatically increased, but few of them cover breakthrough technology. Also, there were few international standards based on Chinese technologies. In short, China is still in the stage of imitating innovation, in which the majority of R&D activities are focused on making incremental progress on foreign core technologies and platforms. To make significant technology breakthrough, we must further alter our way of thinking, reform research mechanisms, foster environments to promote innovation, and encourage the spirit of striving and exploring.

To further demonstrate the status of the information industry in China, global competitiveness rankings from 66 countries and regions published by the Economist Intelligence Unit and supported by the International Business Software Alliance are shown in Table 2-1. Several of the Table's items are related to the research surroundings and ability. China (not include Taiwan, Hong Kong and Macau) get a score of 27.6, ranking $50^{th}$ in the world behind Brazil (31.0, $43^{rd}$), Mexico (30.7, $44^{th}$), India (28.9, $48^{th}$) and Russia (27.7, $49^{th}$). It should be noted that Taiwan, China is ranked second on the list, yet was sixth in 2007. The positions of Sweden and Canada rose by three respectively. The table shows that China is scored low in IT Infrastructure and R&D conditions, partially because of its large population. But the R&D condition is scored so low (1.7) that we really have to rethink the situation seriously.

Table 2-1 Global IT competitiveness rankings for countries and regions (Sample)

| Rank | Country and region | Total score | Enterprise conditions | IT infra-structure | Human capital | Legal en-vironment | R&D con-ditions | Support to IT |
|---|---|---|---|---|---|---|---|---|
| | Ratio | | 0.1 | 0.2 | 0.2 | 0.1 | 0.3 | 0.2 |
| 1 | USA | 74.6 | 98.0 | 89.2 | 94.5 | 92.0 | 23.7 | 86.4 |
| 2 | Taiwan, China | 69.2 | 87.6 | 52.0 | 73.1 | 70.0 | 74.3 | 65.3 |
| 8 | South Korea | 64.1 | 81.3 | 49.3 | 74.0 | 67.0 | 59.9 | 63.9 |
| 43 | Brazil | 31.0 | 66.0 | 13.4 | 38.6 | 46.0 | 1.0 | 61.3 |
| 48 | India | 28.9 | 59.3 | 1.3 | 48.8 | 47.0 | 0.6 | 54.0 |
| 49 | Russia | 27.7 | 46.9 | 10.6 | 55.5 | 38.5 | 1.9 | 36.6 |
| 50 | China* | 27.6 | 46.9 | 5.2 | 46.6 | 59.5 | 1.7 | 41.1 |

*Not including Taiwan, Hong Kong and Macau

## 2.2 Problems and Challenges in the Development of IS&T in China

With the growing IT industry and relatively weak R&D infrastructure, the No.1 challenging task of China is to transfer from an imitator to an innovator. In the process China has to solve problems such as the limited social recognition about the significance of IT, widening digital divide, lacking of long-term foresight and leap-forward development, inadequate understanding about the penetrability of IT, etc.

**(1) Social recognition about the significance of IT.** Quite a part of public and decision makers do not have an adequate understanding about the significant functions of IT. Their way of thinking still remains in the traditional industry age. Huge effort is needed to raise the social consciousness on IT, to change old ways of thinking, to develop IS&T, and to accelerate the process of the combination of information and industry in order to create fundamental changes in modes of production, living, and consumption.

**(2) Digital divide.** In the process of IT development, there has been a trend of the widening of the digital divide. The penetration of PC in urban areas is more than 10 time higher than that in rural areas. The problem of "the last kilometer" in rural areas has not yet been resolved. It will be a long term challenging task to decrease and eventually eliminate the digital divide in China.

**(3) Pursuant and leap-forward development.** In the past several years, much effort was devoted to develop critical common technologies in IT sector, but the overall organization was not as efficient as it should be. There is a gap of 3–4 years (about two generations) between China and the leading countries in critical common technologies in this area. It took 6–7 years to decrease that gap from 20–30 years to 3–4 years. It is impossible to eliminate the current gap of 3–4 years by imitating and following the pioneers. In the coming decades, the development of IS&T in China will be contingent on whether the country continues to imitate or adopts the way of innovative and leap-forward development.

**(4) Long term foresight.** During the past several decades, the biggest lesson in IS&T development in China was that being over-eager for quick success and instant results. Lacking foresight judgment leads to the loss of the opportunities to upgrade China's technology. We seldom carry out research for the new generation technology 5–10 years ahead. Typically every probable direction is listed out in various plans, which are not necessarily able to gain adequate investment, so the end result is poor. R&D funding support for the new generation technologies is not consistent. With hesitation in decision making, windows of opportunity are often missed.

**(5) Inadequate understanding about the penetrability of IT.** We did not recognize the penetrability of IT well enough. Till now, computer science and emulation technology have not been given a proper position. There is not enough emphasis on interdisciplinary research between IS&T and other disciplines.

All in all, in the past 30 years of the reform and opening up, the mode of development in IS&T in China was focused on learning, imitating and following advanced countries. This is one of the main reasons for our current heavy dependence upon foreign technology of Chinese IS&T. In the next 40 years, innovation must be given more importance in comparison with learning and following, gradually become the mainstream of our developing strategy. We should also recognize two historical possibilities: By 2050, China's GDP will top the world and so will IT market and users in China possibly. To meet the relevant long-term demands, we should shift from being a learner to an innovator in IS&T.

## 2.3 Demands for IS&T in China's Economic and Social Development

From a macro perspective, China's economy and society development demands the IS&T to benefit all people, to enable the economy development along a new path to industrialization, to construct a solid foundation for the sustainable and accelerated development of the country as well as the IT industry, to accelerate the development of science and technology, to support the construction of a harmonious society, to safeguard the information security of the country, and so on (Figure 2-1).

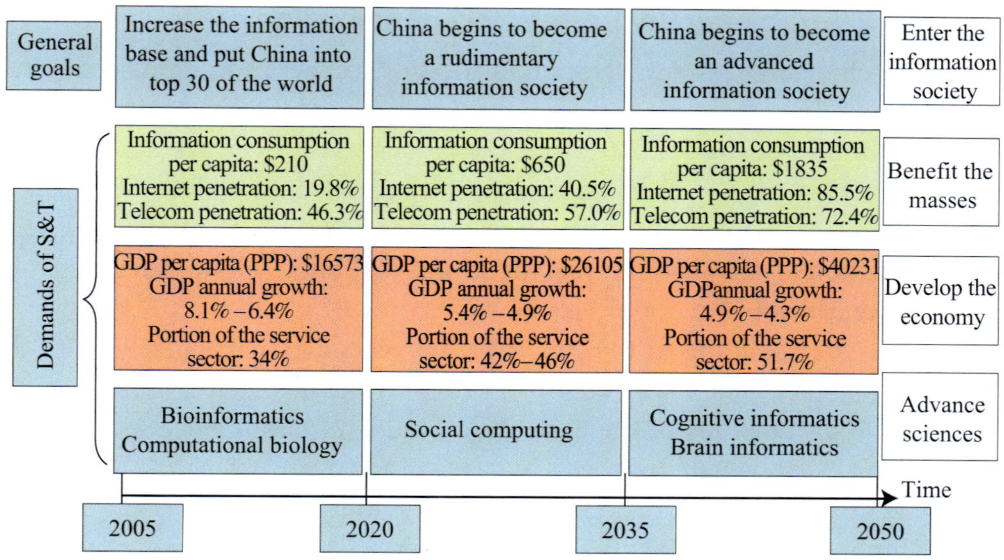


Figure 2-1 Chinese economic and social development demands for IS&T
(Source: Chinese Modernization Report 2006)

### 2.3.1 Benefit the masses

It should be recognized that it is a long-term, difficult task to realize mass adoption of IT. Here "mass adoption" implies that the way of using IT would be as easy as that of using electricity technology—all you have to do is just turning the switch on. It has a long road in front of us to realize this goal. It is not a process to simply expand the usage of the existing technologies. Many of them would be eliminated and new ones invented.

In this process, we must strive to eliminate the digital divide. Rural population and migrant rural workers should be able to get the opportunities to access information resources. These demands generate a new grand challenge to Chinese IT scientists and engineers. We need to develop a new information economics applicable to China's actual condition, establish an IT infrastructure different from that in developed countries with low cost and high efficiency. We should do our best to reduce the total cost of our informatization to benefit the masses.

In the past 50 years, the emphasis was focused on high performance, general purpose, and mass production in the development of IT, while the factors of low cost, ease of use, reliability, safety, and personalization were relatively ignored. Typically the needs of a user will continuously advance from the basic demand to expertise and personalized requirements. Currently, most users' requirements are still focused on the basic functions. In the future IT products have to meet personalized needs of hundreds of millions of users. Computing for the masses should be a strategic direction that needs to persist for a long time.

### 2.3.2 Enable a new path to industrialization

The changes in the development mode and the upgrade of economic structure in China generate tremendous demands for IS&T. Currently China is in the middle of the industrialization process. However, we cannot delay informatization till the industrialization process is completed. We have to choose a new path of industrialization with the feature of integrating with informatization. Thus in some way, the future of China will be determined by how well we develop and popularize IT.

Since the founding of the PRC over 60 years ago, particularly from the reform and opening up, the structure of economy has changed dramatically. Up to 2000, the proportions of the agriculture, industry and service sectors were 15.9 percent, 50.9 percent and 33.2 percent, respectively. By 2020, they will be 11.5 percent, 50.2 percent and 38.4 percent. The industry will keep the position of the number one sector till 2038. Then, the service sector will replace industry with a percentage of 46.0 compared with 45.9 of industry. The share of the service sector will reach 51.70 percent by 2050. (Source: Research Office of the State Council, *New trends of the Chinese economy in the 21st century*)

According to current and planned developing modes, the process of the integration of informatization and industrialization will take a long time. The growth of the service sector will be very slow. The percentage of the service sector will not surpass 50 percent till 2050. In contrast, the service sector of the USA reached 50 percent of GDP in the 1950s. In terms of the structure of the economy, China has lagged behind the developed countries by about 100 years. Following in the footsteps of developed countries, China can not attain the current level of developed countries until the year of 2050. In addition, this kind of mode is accompanied with tremendous consumption of resources and energy and unaffordable pressure on the environment. We have to develop IT to change our modes of living and consumption. The development of the service sector, consisting of finance, logistic, education, healthcare, and entertainment etc., is based on the application of IT, and leads us to a new path of development. Along it we will not only develop the economy in a more scientific way but also guide consuming to expand the portion of IT consumption.

In the new path to industrialization, industry software is of critical importance as the carrier of the integration of informatization and industrialization. Most industry software in China is imported from foreign countries nowadays. There would be no modern manufacturing without a breakthrough in industry software. Thus advanced basic research on modeling, simulation, and industry software should be highly emphasized.

### 2.3.3 Support sustainable development

In the next several decades, China can not realize informatization detaching from traditional industry like some countries with a single-industry economy, nor follow the example of developed countries that begin their informatization only after the completion of their industrialization. The reason is that China's conditions of natural resources, energy supply, population, and environment, which are different from these countries, do not allow it to adopt such development modes. China can not allow its IT and industry to be controlled by outside forces nor depend on the platforms of other countries to develop its industry for long term. The way to realize a sustainable and dependable development of the IT industry, which plays a key role in the national economy and society, is still to be explored.

Currently, IT is not "green". Instead it is an industry with high energy consumption and pollution. Only by becoming a sustainable industry itself can IT support the sustainable socio-economy development. It is a grand challenge to reduce the power consumption, emission and pollution of IT products for the coming decades.

A report from the Chinese Computer Industry Association shows that in the end of 2004 the total number of personal computers in China was 52.99 million with an average total annual power consumption of 20 billion kWh, equal to half of the annual output of the Three Gorges hydropower plant. According to statistics from IDC, the electricity cost of servers in China was 1.9 billion dollars in 2007. The power consumption of China Telecom was more than 20 billion kWh, with an annual cost of more than 10 billion *yuan*. With the development of the Internet and growing number of computers, power costs will increase quickly and consistently. Based on the power consumption level of today's semiconductor devices, if the power consumption increases linearly with the growth of computing capacity, the existing power plants in China will not be able to support a Zettaflops ($10^{21}$ flops) computer, which will become reality most likely around 2030. Even based on the most optimistic estimate that the energy efficiency of chips and information systems increase 50 percent every year, the cost of power consumption alone of a supercomputer after 10 years will be about 1 million *yuan per day*!

The efficiency of value creation of information systems has much room for improvement. According to various international statistics, the hardware utilization ratio of today's PCs is only 1–5 percent, less than 5 percent of the functions of application software are utilized, and the utilization ratio of datacenter servers is only 10 percent. Of the total enterprise IT cost, 70 percent is used in the infrastructure investment and maintenance, and only less than 30 percent is for business innovations. The efficiency problem will become more severe as hardware parallelism increases.

Sustainable development of IT requires not only energy saving, but also raising the efficiency of the allocation of production factors, reducing waste, and changing traditional product structure with the help of IT and its applications. IT systems can also be used for better utilization and exploitation of resources, and transform once unusable resources into useful ones.

### 2.3.4 Accelerate the advance of science and technology

Since the beginning of the 21st century, the penetration of IS&T into other disciplines of science and technology is unprecedented. More and more experts in the world now hold an opinion that information science is changing into a mathematics of new form for other disciplines such as natural sciences, engineering sciences, and social sciences, i.e., becoming the queen and servant of other sciences, just like mathematics. The Internet is making information science into a natural science as well as a social science. IS&T should be considered not only as a tool but also as a fundamental ability, a valuable resource and a fundamental thinking style. In the next 40 years, IS&T will infiltrate into domains of various natural and social sciences more and more deeply, play important roles in all sectors of economy and social life to meet the growing material and cultural demands of the public.

This era of profound changing, which may last for several decades, just began. We should recognize this historical opportunity.

Since the reform and opening up, in economy China has found a unique developing path with Chinese characteristics. But in the field of S&T, in some sense, we still remain in the status of follower. We should grasp the opportunity brought by the penetrating process of information science into other sciences, address strategic problems in the development of our country, and find a developing path with Chinese characteristics for IS&T in the 21st century with the heritage of brilliant tradition of Chinese culture.

The penetration of IS&T can not only promote the development of bioinformatics, computational biology, brain science, cognitive science and nano science, but also address those tremendous problems related to energy, healthcare, education, employment, environmental protection, and global climate change.

### 2.3.5 Ensure the national information security and social harmony

In the age of knowledge economy, information is one of the most important strategic resources for a country. The Iraq war shows us the importance of *the domination of the information*. The overall security of the Internet and information systems in China is not entirely positive. Hacking and computer crimes, the spread of Internet viruses and Internet terrorist activities are real threat. According to governmental statistics, 80 percent of web sites in China contain hidden dangers and 20 percent of web sites in China have serious safety problems.

Cyberspace is the fourth strategic frontier besides the sea, the land and the sky. The safety of cyberspace is of paramount importance. Network security concerns have been expanded beyond Internet to include other fundamental networks such as telecommunication networks and broadcasting networks. It becomes more and more difficult to ensure the security of these networks. Internet viruses and hackers are new threats to these networks. Research on information management and inspection are areas that need much attention in the development of IS&T. Network survivability and vulnerability and network trust are important research topics. The traditional concept of computer security has transferred to the information security based on the trustworthy computing. The strategic aim of information safety is to construct *a trustworthy* information infrastructure.

We have to keep balance between openness and safety. On one hand we should uphold the network openness principle to enable IT to benefit the masses; on the other hand we should strive to construct a safe and harmonious network, and strengthen honesty and self-discipline, protect people's privacy, eliminate annoyance from junk email and harmful information by the help of network security monitoring technology, and ensure a healthy development of networks and information systems.

## 2.4 Opportunities for Development of IS&T in China

Up to 2050, there will be three major opportunities for the development of IS&T in China: information science will see important breakthroughs in the first half of 21$^{st}$ century; China IT market will top the world; China's innovation ecosystem will see substantial improvement. If we can grasp the opportunities and create a long term plan, China has the potential to become a pioneer in IS&T.

Now we are in the stage of an information revolution, which will have a deeper effect on human life than the industrial revolution or the agriculture revolution. It may take about 100 years to fully understand this information revolution. Some experts think that as an analog to physics, computer science is still in a pre-Newton stage, and the work we do now is like the observations work done by Galileo. Therefore, in the long term, the development of information science and technology has ample room for innovation.

As a whole, the process of development and use of IT in China is still in the initial stage, and has large room for further development. According to data from the Ministry of Industry and Information Technology, one of China's goals is to make the IT expenditure per capita of 2050 to reach the level of 2000 in developed countries. Based on these data, China's IT market could grow to 1–2 trillion dollars by 2050, becoming the largest in the world in terms of both scale and users. The information market today is only 10 percent of that in 2050, most of which is yet to be explored. The monopoly market we see today is not necessarily the same in the future. Innovation for new IT markets holds the largest innovation opportunity.

The innovation ecosystem in the 21$^{st}$ century will be much improved, marked with the participation of governments, academia, companies, and users. In it all participants can create and realize their value through open source codes or paid products and services. It will be more and more difficult for large companies to monopolize the market. The innovation ecosystem in the 21st century will be like an ecological environment in the natural world. Every member, no matter how big or small, can find value and play an individual role and realize his/hers value. A main motivation to develop the technologies, such as networks, virtualization, service, socialization and intelligence in the beginning of the 21st century is to eliminate the tight coupling phenomena in the information industry, to eliminate dependence and control, and to liberate productivity, thereby to promote integrated innovation, combinatorial innovation and autonomous innovation (innovation that does not need the authorization from others). Before the 20th century, many inventions came from amateur scientists. In the past 100 years, most of the great inventions in S&T were supported by the governments and enterprises. In the 21st century, the barriers for scientific research will be further lowered and amateur scientists will spring up again. Interdisciplinary research will be a trend.

Whether China will be a force in the innovation and competition in the coming decades and be the pioneer of IS&T is a serious question for us to answer in practice. Many foreign colleagues hold an optimistic opinion on China. In 2004, IEEE Spectrum published a survey of IEEE fellows to answer "*which country will be the innovation center in the world in 2015?*" Sixty-five percent of the fellows chose the USA, while the second place went to China with 17 percent. This report illustrates in some sense that China can make bigger progress in the future.

### Information Technology (IT) and Information Communication Technology (ICT): the Chinese and Global Markets

The information market size of a country is an indicator of its level of the IT (and ICT) use in the country. Common international practices use IT consumption data, i.e., IT expenditure on computer and network hardware, software and services, to measure the IT market size. A broader metric is ICT expenditure (Information and Communication Technology Expenditure) including IT expenditure and communication expenditure.

According to *OECD Information Technology Outlook 2008,* the worldwide ICT expenditure was 3.766 trillion US dollars (IT expenditure accounts for 41%) in 2008, while that of China was 327 billion dollars. The table below forecasts the Chinese IT market in 2040 and compares to actual data in 2000 and 2008. (For a more detailed explanation of this table please refer to Chapter 7)

| 2040 forecasts vs. 2000 and 2008 data | GDP per capita | IT market (trillion dollars) | Portion of IT market in GDP | Annual growth of IT market | IT users (million) | Annual IT expenditure peruser | Annual IT expenditure per capita |
|---|---|---|---|---|---|---|---|
| 2000 (U.S.A.) | $34,600 | 0.4 | 4.10% | | 161 | $2,484 | $1,418 |
| 2000 (China) | $854 | 0.026 | 2.40% | 25.0% | 22.5 | $1,154 | $21 |
| 2008 | $3,300 | 0.11 | 2.55% | 12.7% | 270 | $415 | $85 |
| 2040 Conservative estimate | $14,000 | 0.5 | 2.49% | 4.75% | 1,200 | $415 | $332 |
| 2040 Neutral estimate | $14,000 | 1.1 | 5.14% | 7.33% | 1,200 | $900 | $727 |
| 2040 Optimistic estimate | $30,000 | 2 | 4.40% | 9.38% | 1,200 | $1,651 | $1,321 |

(Data of 2000 and 2008 are of current price and come from Chinese National Bureau of Statistics, Ministry of Industry and Information Technology, CNNIC, ITU, *Report on Chinese Modernization 2006*. Data of 2040 uses the price of year 2008; and the data for China are in terms of exchange rate, not PPP.)

# 3 Strategic Targets for the IS&T Development

## 3.1 Overall Objective for the Development of IS&T—To Realize an Information Society

From 2010 to 2050, the general target of IS&T in China is to grasp the opportunities brought by the coming transformative changes of information technology, improve abilities for innovation and sustainable development, and enable China to become a full-fledged information society. In this society, most people will be information users; information will be the most important resource for the economy and society; and the informatization level will be close to that of developed countries.

### 3.1.1 Expectations for the information society

Information society is seen as the successor to industry society. Developed countries have been practicing principles of information societies since the late 20$^{th}$ century. Currently, China is in the middle of the industry stage. In the following several decades, following a new path to industrialization, China will gradually become an information society through the integration of informatization and industrialization.

The following are the expectation for the information society by 2050:

(1) Information will be the most important strategic resource. Information is not only a productivity factor but the most active one. The production, processing and handling of information will be the basis for social wealth creation. Knowledge and information services providing will become a leading industry.

(2) Informatization will become the critical factor affecting the economic efficiency and process, such as manufacturing, logistics, circulation, management and consumption. Informatization and networks become defining characteristics of society.

(3) Ubiquitous networks will integrate the networks in space and on ground and access networks, to realize people to people, machine to machine and people to machine communications anytime and anywhere. The ubiquitous networks will support dependable and low cost services.

(4) An integrated and intelligent information network infrastructure will be offered to the public. The information networks will be a platform not only for communication and knowledge sharing but also for knowledge processing and knowledge computing. Everyone will be able to consume and access information according to his or her requirements. Every family will have an information "socket" in addition to power sockets. An "information meter" will be the base for charging the use of information and services, similar to the electric meter we are using now.

(5) Thousands of kinds of efficient, reliable, energy saving information devices, most of which do not exist yet and need to be created in the future, will be widely used in various applications. Users will be able to freely choose the functions, the usage mode and the consumption of information as they want. They will also be able to conveniently develop new applications, new functions and interfaces, instead of being pre-determined by manufacturer or service provider, and do so without the need of being approved by anybody. Users will be able to switch service provider as easily as they switch TV channels. Production and consumption will be integrated and accelerate each other's pace of development.

(6) An open and dynamic network society, which consists of many people and many machines, will emerge, to enable interactive innovation and contribution. Computer networks will not only execute pre-deployed algorithms like a computer, but become more and more intelligent by "learning" from usage. The process of generating intelligence by "people to people", "people to machine" and "machine to machine" interaction will eventually evolve to a process of "social intelligence" emerging.

(7) Different from the individual autonomy of an agriculture society and forced order of an industrial society, one characteristic of an information society is autonomous order through cooperation. The industrial society is featured with conflicts between human beings and nature, struggles among people, and spoiling of resources. In information society, the coordinative function of information is emphasized and individual achievement is realized by self-organizing to optimize the position for every social unit.

(8) In the future information societies, devices and systems will be miniaturized and get to be more intelligent. Microchips might be embedded into human body and brain. In daily life, robots will be partners with human beings. Quantum cryptography technology will make the information system secure. Access to information and business processes will be realized through DNA recognition. IT will be a silent, serene and harmonious technology, and like electricity, it will become an invisible technology.

(9) Most of the Chinese population will be Internet users, the digital

divide will be eliminated, and everyone will be guaranteed to have a universal access to information resource at a certain but growing minimal level. The market for information applications will become bigger than the one of IT products. Information service consumption will be much larger than that of software and hardware. IT will help environment protection and become sustainable.

### 3.1.2 Two stages of information society: e-society to a u-society

In terms of the progress of information network technology, the development of the information society can be divided into two phases: e-society and u-society. Before 2020, China is to establish the base for an information society—e-society. Then the task is to transition into a u-society. The "U" has three meanings: Universal, i.e. benefiting all people, User-Oriented and Ubiquitous. The e-society is the primary stage of the information society and the u-society is the advanced.

For China, the goals of the economy in the phase of e-society are: GDP per capita exceeds 6000 US dollars (the level of middle income countries), the human development index is greater than 0.8, the human poverty index is less than 10 percent, the urbanization rate is greater than 50 percent. More goals include developing the modern industrial system; promoting the integration of informatization and industrialization; developing modern service industry based on the Internet; improving e-government; getting 60 percent of companies to have their own websites, raising the rate of online purchasing to 30 percent; developing new media, cultivating a new type of cultural industry with high speed new transmission technologies.

In 2003, the World Summit of Information Society hosted by the UN created ten targets to be achieved by 2015: to connect villages and establish community access points; to connect universities, colleges, secondary schools and primary schools; to connect scientific and research institutions; to connect all public libraries, museums, and archives; to connect health centers and hospitals; to connect all local and central government departments and establish websites and e-mail addresses; to adapt all primary and secondary school curricula to meet the challenges of the information society, taking into account national circumstances; to ensure that the entire world population has access to television and radio services; to encourage the development of content and put in place technical conditions in order to facilitate the presence and use of all world languages on the Internet ; and to ensure that more than half the world's inhabitants have personal use of ICT. The above targets have formalized the requirements of an e-society. Although China is a big country with a substantial population still living in poverty, these targets may be realized by 2020, if we work hard to eliminate the digital divide.

For China, the economic target of u-society is to mark China's GDP top of the world, the economy a healthy and sustainable one. The social targets include realizing harmony between human beings and nature; becoming a resources conservation society; improving the quality of the ecological environment and living space, and entering into a sustainable virtuous circle; connecting anyone and anything from anywhere at anytime, and solving the economic and social problems. The targets of an e-society and a u-society are shown in Table 3-1.

**Table 3-1 Primary and advanced stages of an information society**

| Time | | 2000 | 2020 (e-society) | 2050 (u-society) |
|---|---|---|---|---|
| Technology popularization | Computer penetration | Total number of computers: 15.9 million, penetration: 1.3% | Total number of computers: >0.5 billion, penetration: >50% | Almost everyone has access to information terminals |
| | Network penetration | Internet users 16.90 million, penetration : <2% | Internet users: 0.6 billion Rural Users: 0.3 billion | Information network is as ubiquitous as the power grid; the digital divide is almost eliminated |
| | Ease-of-use | Most people do not know how to use the computer | Most people know how to use the computer | Almost all people know how to use the computer |
| Freedom of users | Freedom of choice | Few choices; difficulty in changing suppliers | More choices; being able to change suppliers | Many choices; being able to change suppliers freely |
| | Freedom of innovation | Need permission from vendors or service providers, confusing and conflicting standards, high barrier | Simpler standards, requiring less permission | Fusion of information producer and consumer, freedom to innovate |
| Communication | Communication between p-p, p-m, m-m | Mostly limited to people-people with geographical restrictions | Wireless communications with little geographical restriction, emerging sensor networks | Two-way seamless links among the physical world, cyberspace and human society |
| Richness of resources | Efficiency and trust | Low efficiency and trust; information silos | Higher trust and efficiency; fewer information silos | Almost unlimited information processing capacity; high security based on quantum cryptography |
| | Flexibility | Numerous and disorderly information resources, scarce service resources and few high quality information resources | Rich service utility resources, the Internet becomes the main source of information | Machine understanding of semantics, wide use of semantic searching and machine translation, industry of data intelligence |
| Ecosystem | Participation of the public | High-tech barrier, vendor domination, openness promoted by Internet | Participation of all people; high-tech myth broken; open standards | Open standards dominate, intellectual property rights are no longer obstacles for innovation |
| | Regulation of the government | Only in communication; anti-monopoly lawsuits in the computer industry | Expanding to networks of computers, emphasis of personal privacy | Harmony among government regulation and personal freedom, personal privacy protected |

## 3.2 Basic Framework of IS&T Development up to 2050

This roadmap study focuses on China's demands for IS&T when it enters information society, relevant goals of IS&T and the selection of the path to the goals. The selection of the path means to identify key scientific problems, key technologies, targets and the time required for realizing these technology breakthroughs.

As IT industry changes constantly, the future of IT, especially for such a long period of several decades, is difficult to predict. Generally speaking, the progresses in science are neither always linear nor incremental but rather involve many disruptive discoveries. It is difficult to identify the strategic task for the development of science and technology in the coming decades accurately based on the current status and general goal of information society. The aim of the roadmap is not to try to create a precise prediction, but rather to make some rational prospective judgments about the IT development goal and the path to it, based on the general rules of S&T development and the strategic demand of China.

The best way to forecast the future is to create it. The development of S&T is not a process independent from the people's efforts. The purposeful activities of human beings will have a direct impact on the future of S&T. Within the next 40 years, we will sail into a sea of IS&T without marked channels for navigation. The roadmap, similar to the most simple navigation techniques like the using of the direction of wind and currents, only tries to offer some understanding about the challenges, trends, and their relations.

To allow China, a country that is currently in the middle stage of industrialization and lags behind the developed countries in informatization, marketization, urbanization and globalization, to enter into information society, efforts in S&T should be focused. The key goals in different developing stages have to be set wisely according to China's national conditions and the inherent rules in the developing of IS&T. Setting these goals is the fundamental task of the roadmap. Figure 3-1 shows the main judgment and trade-off in the roadmap. The red arrows show the paths that we should follow while the blue arrows show the ones to be avoided.

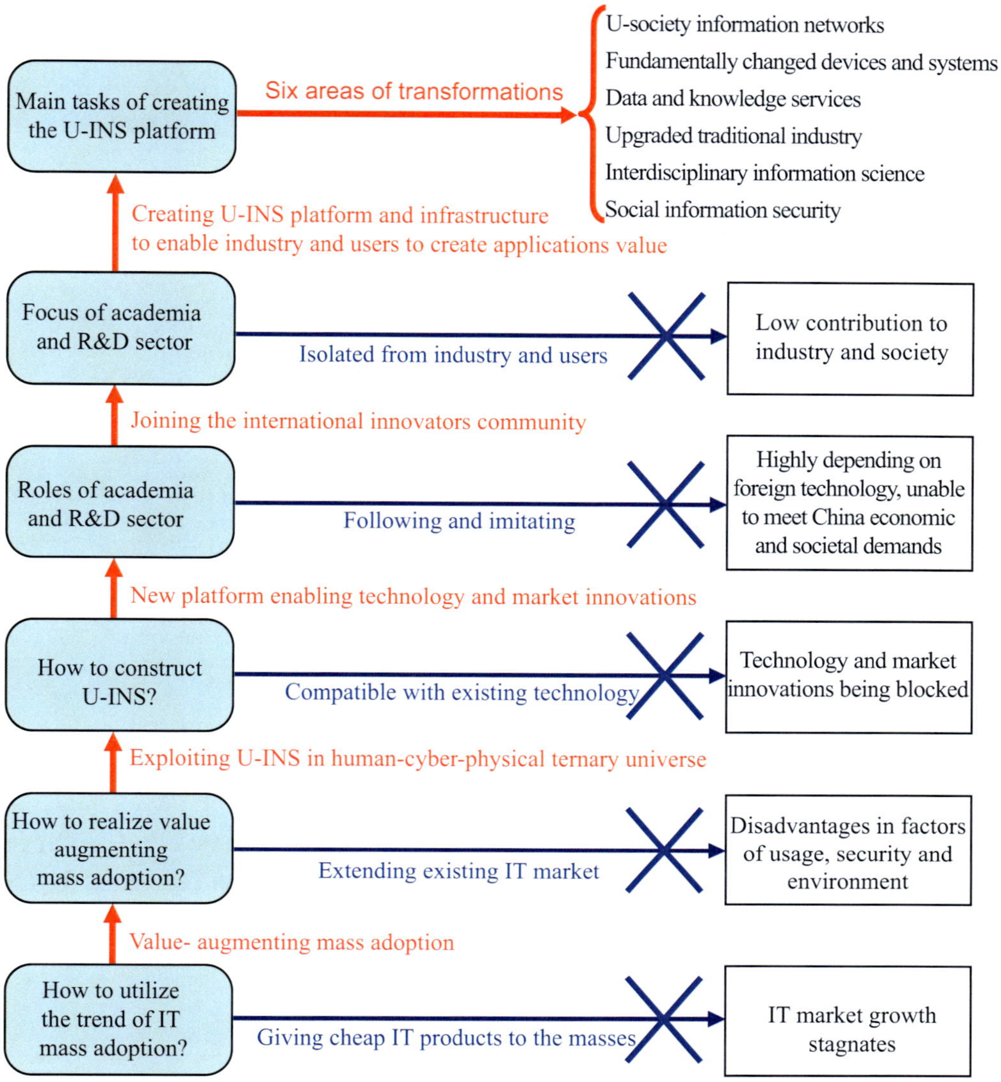


Figure 3-1 Main decisions involved in the roadmap of IS&T development

According to the analysis on the status and trends of IS&T in China and the world, our team affirms that the following six tasks should be focal points for China in the next 10–40 years:

(1) To construct ubiquitous, well-content information networks.

(2) To upgrade devices and systems.

(3) To develop a data and knowledge service industry.

(4) To upgrade traditional industries and realize low-cost informatization.

(5) To develop new information science and interdisciplinary sciences with computing as a connecting bond.

(6) To construct a national social information security system.

Each primary task of the roadmap is comprised of several sub-tasks. The basic framework is shown in Figure 3-2.

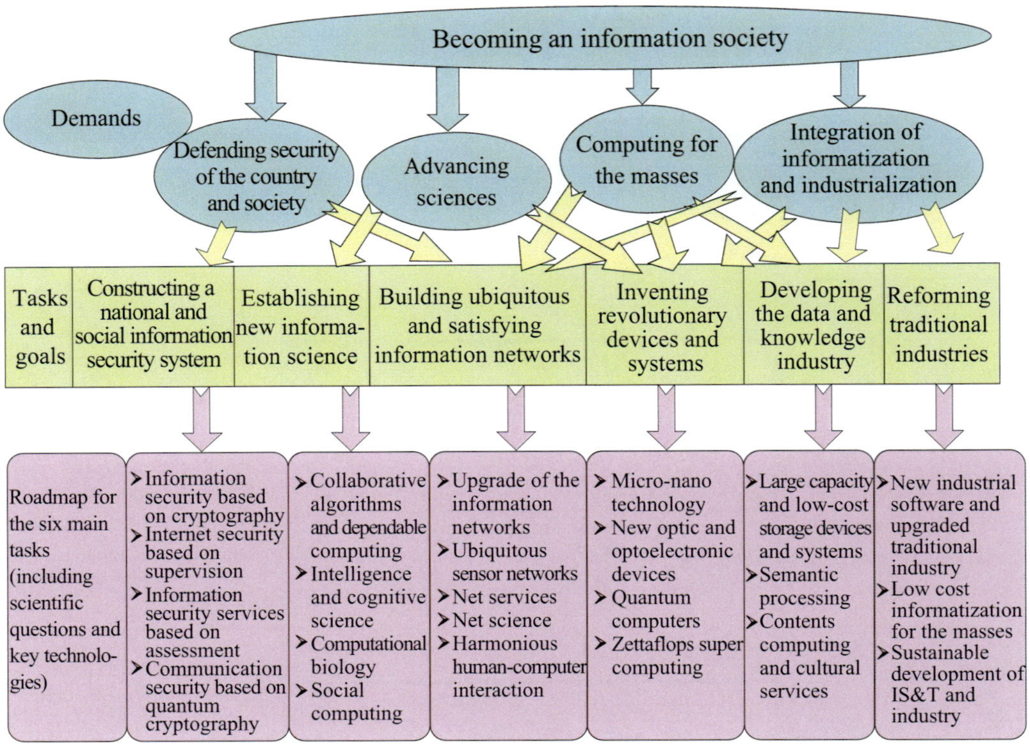


Figure 3-2 Framework for the IS&T development roadmap

These six tasks boil down to an over all goal: **to construct Universal, User-oriented, Ubiquitous Information Network Systems (U-INS). U-INS** includes six components (as shown in Figure 3-3):

(1) Supporting networks and devices and systems of new generations.

(2) Universal and User-oriented systems.

(3) Safe, trustworthy and individualized net service.

(4) Network applications and industry.

(5) Network science and new information science.

(6) National and social information network safety systems.

Tasks 1 to 4 show the four levels from single devices to net applications and network systems. Network science and safety systems are guarantees for network systems and run through every level. They show the demands in China in the first half of the 21$^{st}$ century and the key S&T that is required. Universal, User-oriented, Ubiquitous Information Network Systems should be highly emphasized and developed actively in China in the next 10–40 years.

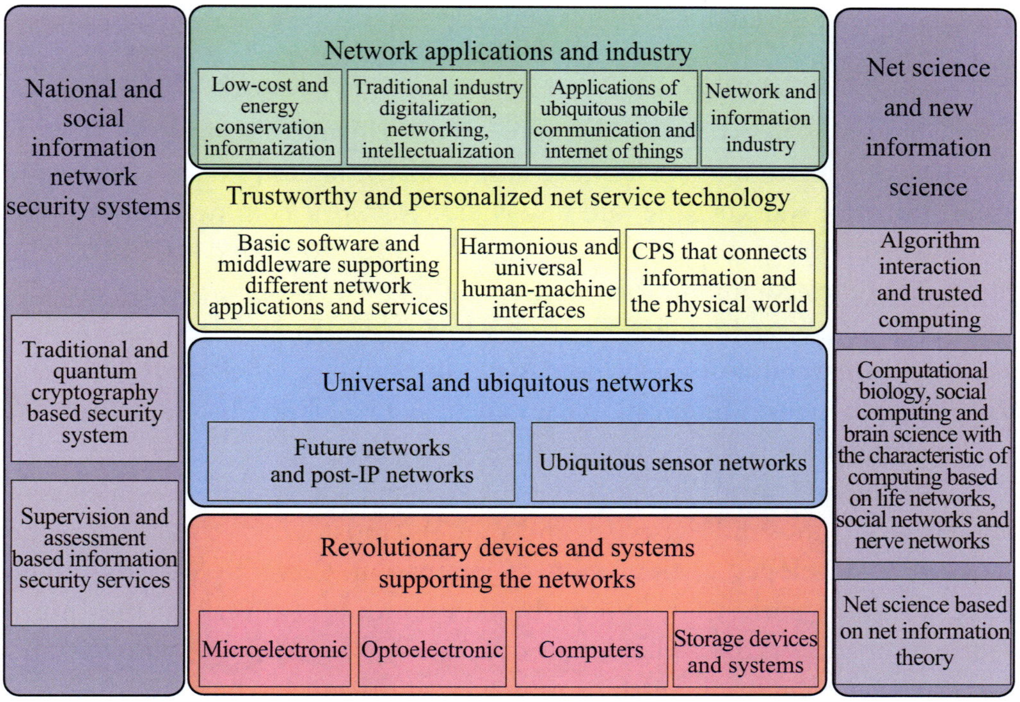


Figure 3-3 Universal, User-oriented, Ubiquitous Information Network Systems (U-INS)

### 3.2.1 Universal

The rise of Internet-based modern services and new media are changing the social, economic and political patterns of human life. In the 21st century, IT is entering the stage of mass adoption. Factors such as low-cost, security and ease-of-use have become more important than performance. A key task is the development of low-cost, energy efficient information terminal devices, servers and information services. We should strive to decrease the total cost of informatization in China, and provide network access to all people at low cost.

### 3.2.2 User-oriented, user-centric

With the rapid development of the Internet and Web 2.0, service modes are changing from enterprise-centric to user-centric. In the past several decades, the market for information has been a seller's market; manufacturers and service providers have too much control. But the situation that the users are forced to upgrade their devices and accept the unwanted services is changing. The U-INS will further develop user-centric networks.

In the future networks, the user will not only be a consumer of predetermined services. Future networks will provide services, which are composed from smaller services, according to the demands of users. Future networks will be not only for communication and information sharing but also for service in nature. Future networks will not be upgraded by people anymore, but be adaptive for self-tuning and self-optimization, be able to dynamically adjust contents and management to minimize service provider's interference. The user will be not only an information service consumer but also a creator and supplier of information contents and services.

### 3.2.3 Ubiquitous

Ubiquitous networks can be integrated with space, ground, and access networks, to realize human-human, human-machine and machine-machine communications at anytime and anywhere. Network communications will become pervasive, stable and low-cost services. We need to realize the integration of telecommunication networks, the broadcasting networks and the Internet to utilize satellites, wireless and wired networks for access service. We also need to develop intelligent miniaturized terminals: efficient, trustworthy, and power-saving communication equipments and facilities, including human-machine interaction systems based on neural engineering.

The future information society could contain trillions of such intelligent devices with thousands of types and models, most of which are yet to be invented or developed as the results of innovation. Communications among devices will be more than human-human communications. In the future information society, the combination of IT and nano technology, biology and cognitive technology will enhance miniaturization and intelligence of devices and systems. Micro chips can be embedded into human body, even the brain, to enable the monitoring of inner body functions and the circulatory system.

Table 3-2 shows the characteristics of the universal and ubiquitous information networks by 2050. Four specific goals of China set for 2050 are:

(1) Goal for popularization: The penetration of information networks in China will reach 80 percent, and China will top the world in terms of the number of information users.

(2) Goal for universal coverage: The digital divide will be eliminated. Hundreds of million people will be lifted from information poverty. The information consumption per capita in China will reach the level of 2000 in developed countries. Information thinking will be popularized.

(3) Goal for market: The scale of IT market in China will reach 2 trillion US dollars, and become one of the largest in the world.

(4) Goal for innovation and market openness: The IT market of China will get rid of the domination from oligarchic enterprises and the ratio of foreign technology dependence will be reduced to less than 20 percent, so that entrepreneurship will have enough space to develop.

Table 3-2 Characteristics and targets of Chinese universal and ubiquitous information networks by 2050

| | | Around 2020 | Around 2035 | Around 2050 |
|---|---|---|---|---|
| IT penetration | Terminal penetration | More than 500 million computers, Portion of new type terminals > 50% | Penetration of ubiquitous terminals > 80% | Almost everyone has an information terminal |
| | User penetration | more than 600 million users, half in rural areas | One billion users; sensor networks will be popularized in both urban and rural areas | Information networks will be as popular as the power grid, and information thinking will be popularized |
| Networks capability | Wired networks | LAN bandwidth > 100 Gbps, user access speed > 1 Gbps | Post-IP networks, metropolitan area quantum cryptography communications system | Network bandwidth on demand, Practical communication networks based on quantum cryptography |
| | Wireless networks | 100 Mbps bandwidth for user, well developed mobile Internet | Merging space, air, ground and submarine network communication | Intelligent wireless communication |
| | Sensor networks | Penetrating into logistics, medical monitoring, environmental protection and disaster prevention | hundreds of billions of sensors deployed | Ubiquitous smart dust |
| Information services capability | Resources on server side | 85 million domain names, 14 million websites, 60 million servers | Ubiquitous Internet expertise services, rich information service sources | Personalized and intelligent services become mainstream |
| | Information contents | 330 billion web pages in Chinese | 10% of worldwide web information will be in Chinese | Sufficient for all personalized demands |
| | Scale and quality of the information industry | Annual IT sales of more than 1500 billion *yuan*; Enhanced innovation capability | IT platforms free from monopoly; a competitive IT industry with zero growth of energy consumption and emission | A data and knowledge industry becomes one of the pillar industries |
| Dependence on foreign technology | | <30% | <25% | <20% |

The process of Chinese informatization to 2050 can be divided into three phases.

The first phase (2005–2020): **Strengthening the informatization base**. By 2020, the gap between IT applications in China and that of moderately developed countries will decrease from over ten years to seven years. China's ranking in Networked Readiness Index will move from the present $50^{th}$–$60^{th}$ to $30^{th}$ in the world and $5^{th}$ in Asia. China's eastern coastal areas will enter the primary stage of information society.

By 2020, China's information industry will mainly be supported by domestic companies and self-created intellectual property rights, with solid equipment capability. A sustainable model for information industry development with low cost and high efficiency will take shape. The value added of ICT products and services will account for more than 9 percent of GDP. A national informatization management system of regulations and laws and an information sharing system among different sectors will be constructed. IT will penetrate into all economic sectors. Traditional industries will be upgraded based on IT in large-scale. Digital divide between rural and urban areas, as well as between the eastern and western parts of China, will be reduced. Informatization will become an important instrument to improve living standards, raise the level and quality of education and build a harmonious society. A social information service system for people of all sections and groups will come into being. The informatization level of public health, employment security and social management will be greatly lifted. E-government will be widely applied, so that the macroeconomic management information environment be changed.

By 2020, desktop operating systems and office software, servers and basic communications and network equipments based on intellectual property created in China or open source technology will account for more than 50 percent of the Chinese market to construct the solid foundation for cost reducing of informatization.

The second phase (2020–2035): **All-round developing of informatization.** By 2035, the level of IT applications will witness a great leap. The gap of informatization level in the economy and society between China and moderately developed countries will be reduced to about two to three years. China's ranking in Networked Readiness Index will move to 20^th^ in the world and 3^rd^ in Asia. In the eastern coastal areas, the informatization level will approach the level of moderately developed countries.

By 2035, the share of the products and services based on China-created intellectual property and open technology will get to more than 50 percent in China IT market. The cost of informatization will be greatly reduced. The leading service providers and manufacturers will be top multinational corporations in the world. Long-acting service mechanisms will be financed based on the core of the universal service fund to provide a solid basis for China to develop a more equal ICT environment to cover its rural area and west part and to extend the content of universal service. The value added of the ICT industry will account for more than 12 percent of GDP. A national informatization management system of regulations and laws and an information sharing system among different sectors will be well established.

The third phase (2035–2050): **Achieving highly developed informatization**. Informatization in the economy and society will be highly developed, with an overall level equal to or higher than that of the moderately developed countries. China will be ranked top 10 in terms of Networked Readiness Index, become a moderately developed information society and the informatization level will catch up with the level of developed countries in the beginning of the 21^st^ century. Eastern regions in China will get into advanced information society.

By 2050, almost anyone in China will have information access devices and the Internet penetration will increase to over 80 percent. Computing power will be provided on personalized demand. Quantum cryptography will make information systems secure and trustworthy. Government regulating and personal freedom will be balanced with harmony. Personal privacy will be highly protected. An informatization pattern for modern society will be established with the leaping development of information applications. China's IT industry will have strong international competitiveness. People become educated citizens in information society. IT will continuously and comprehensively penetrate into various sectors in economy and impact on various aspect of peoples' life, and become a platform for elevating living standards and ensuring social harmony.

### Nine Characteristics of the Roadmap to 2050 for IS&T in China

The overall goal of IS&T in China is to construct a universal and ubiquitous information network systems (U-INS) with the following nine characteristics:

- **Universal:** Information networks penetration will be more than 80 percent.
- **Value augmenting:** China's IT market will grow 10-20 times, lifting 1.2 billion people from information poverty.
- **Ubiquitous:** Information networks will be pervasive and ubiquitous, and informational thinking will infiltrate into various disciplines.
- **Low costs:** People will enjoy free basic services and many low-cost information services in addition.
- **Sustainable:** As IT market grows, its energy/resource consumption and emission will not increase.
- **Safe:** People will see a safe and harmonious ternary universe of human-computer-thing and a civilized cyberspace.
- **Transformative:** The above objectives will be achieved through transformative innovations, not just incremental development.
- **Innovative:** China will change from an imitator to an innovator in IT, reducing foreign technology dependence to less than 20 percent and making its contribution to the world.
- **Platform:** The academic community will create the open U-INS platform, while companies and people will create value products and services on it.

## 3.3 Challenges and Issues in IS&T up to 2050

To realize these targets, especially to realize a universal and ubiquitous information network, two conceptions have to be changed:

### 3.3.1 The narrow mindset that ICT is merely a tool should be abandoned

Based on experiences and lessons in China's development of S&T, whether the cognition is right or not can determine whether we can perceive the future and thus realize a revolution in S&T to change from continuing to be an imitator to becoming a creator. To realize the aims that everyone has access to information terminals and almost any networked device is an information terminal, we should strive to spread computational thinking and information thinking to every person, to make all companies and people be users of IS&T and be able to create value with IT.

Computers and the Internet have long been regarded as high-tech tools. Computer science has also been viewed as a professional tools discipline. This kind of social cognition created a negative impact. "High-tech" is always associated with high threshold and high cost for using in people's mind, and "tools" generally implies an "assistant" instead of a pillar industry creating value for economy and society. This narrow mindset is harmful for the convergence of IS&T and other industries, and has a negative impact on IT popularization.

Throughout history, many general-purpose technologies have surpassed the obstacle of "tools image". Taking reading and writing, the electricity sector and the petroleum industry as examples, people generally do not view them as merely a tool. We should understand that the world has already begun to enter into an information society. In the 21st century, IT provides not only tools, but also basic ability, resource and capital. Production of data and knowledge is an important part of the sustainable economy just like mining and manufacturing.

In the past 10 years, the international information scientific community has thought much about how to overcome the obstacle caused by "tools image". Work in basic research projects, exploitation of forefront technologies and discipline restructuring has been planned and implemented to address this issue. We also see that in the global IT market, the mode of information usage is changing from mere consumption to user creation, use as production and use as investment. The perspective on IT must be changed from viewing IT as mere a tool to as essential resources, fundamental abilities and value-augmenting assets. This is a basic task for the development of IS&T and a basis for addressing other challenges.

### 3.3.2 The traditional modality of man-machine symbiosis has to be transformed to the idea of man-computer-thing ternary universe

Present information systems are mostly based on the man-machine symbiosis modality appeared over 40 years ago. The symbiotic system is composed of the human (user) and the machine (computers and other IT systems). The work is divided in the way that the human is responsible for instinctive and intuitive jobs and the computer is responsible for predetermined and mechanical ones. The human normally is in charge of setting the goal and motivations, making judgment and decisions. The machine is designated to tackle the details and execute according to the prescribed procedure. This division of labor matches well with Turing's algorithmic computational model, and drove the development of IS&T for more than 40 years.

The computing world we see today is different from the human-machine symbiosis system. We are in a ternary universe composed of the cyberspace, the physical world and the human society (shown in Figure 3-4). The ternary universe is different from the human-machine symbiotic system consisting of one computer for one person and a clear division of labor. Instead it is a dynamic and open network society composed of numerous people, computers and things. This transition makes information science change substantially, to a science for studying information processing in the ternary universe mentioned above.

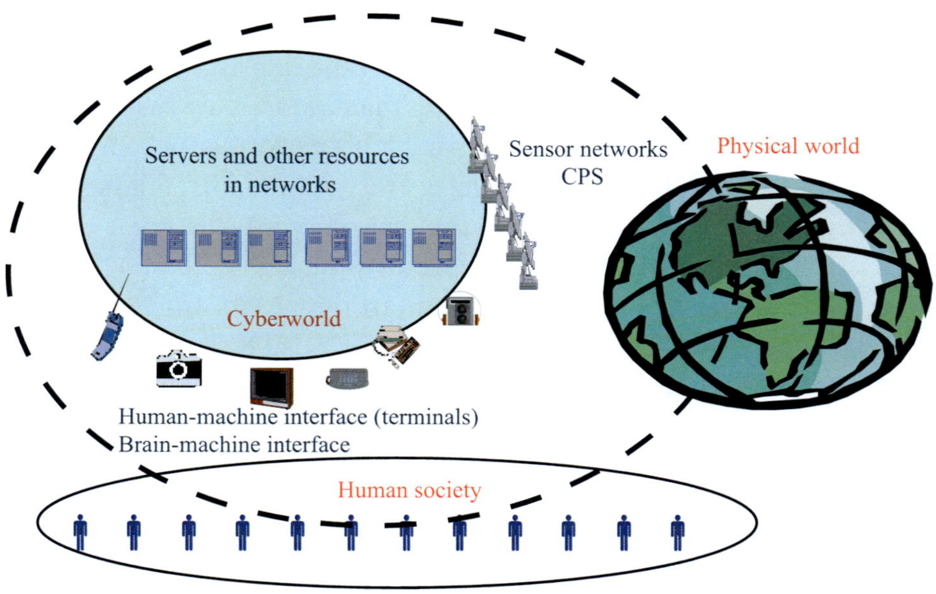


Figure 3-4 The ternary universe composed of the cyber world, the physical world and the human society

Many issues are to be resolved in science, engineering and discipline construction for the human-computer-thing ternary universe. We still can not answer the following basic questions: Can the World Wide Web be regarded as an abstract computer? What is the computability of the World Wide Web? What is the instruction set of an Internet-of-things computer? What is the definition of computation in the human-computer-thing ternary universe? Is it a kind of Turing computation? How can we define the time and space complexity in the human-computer-thing ternary universe? Do we need other basic complexity metrics?

Nowadays, the cyber world is merging into the physical world and the human society (as shown in the dashed line in Figure 3-4). IT research could be artificially made overly complicated, when it is separated from the related physical processes and the human cognitive processes. An alternative is to combine IT research with the practical problems in reality. The ternary universe brings changes in usage modes and workloads. For example, temporal locality and spatial locality are important characteristics of traditional computational workloads, which enable us to decrease the costs of information equipment and information applications. We do not yet understand what locality principles exist in the ternary universe. Many application systems have been developed in areas such as data centers, Internet services, networks and social computing, etc. But the application systems are mostly developed in an ad hoc way, lacking the guidance from new IS&T. Consequently, the cost of these application systems is quite high, and few of them are successful.

In the development of IS&T, the following six essential issues need to be addressed.

**1. Difficulties in hardware development**

In the past 40 years, IT hardware has been largely dependent on the progress following Moore's Law: the quantity of the transistors in a chip doubles every 18 months. Exponential growth faster than Moore's law has been observed in memory density, network bandwidth and high-performance computer speed. For example, network bandwidth doubles every six months, following Gilder's Law. Today the CMOS semiconductor technology faces great challenges. Moore's Law has guided the semiconductor industry for half a century like a beacon. This beacon will begin to dim around 2020. Integrated circuits, Internet, high-performance computer and memory technology will meet a technology wall then, if no breakthrough emerging. The problems of cost and power consumption are going to be very serious. The cost for the constructing a 32-nanometer CMOS foundry reached 4.2 billion dollars. In a few years, the cost for an 18-nanometer production line may exceed 10 billion dollars likely.

Before 2050, we must change the habitual thinking about Moore's Law, and we have to find new important rules for the development of IS&T and seek out other paths of value growth for the information market. One way of thinking is to view hardware as the first industry in the information market similar to agriculture in economy, software as the second industry similar to industry and service as the third industry. The share of the latter two should be significantly increased. Network effect could become more important than Moore's Law. Industrial consortia will create the roadmaps for key software platforms and service technology, much like today's work done by the ITRS for roadmap of semiconductor industry.

**2. Difficulties in large-scale parallelism and programming the ternary system**

A significant change brought about by the ternary universe is large-scale parallelism. The network characteristic of the ternary universe implies large-scale parallel workloads. Information software and services will use many parallel, distributed and decentralized programs. There may be one million cooperating parts in an information solution system. All hardware platforms will also be parallel. Hardware parallelism of a server system could reach $10^9$ by 2020. Client computers and mobile phones will also adopt multi-core chips.

Although both workloads and hardware present a parallelism trend, we have accumulated little knowledge and technology of parallel programming. People start to realize that parallel programming is as difficult as artificial intelligence. Current codes running in client devices are mostly serial programs. In server side, only a few applications for scientific computing can use parallelism up to thousands. What we need is a programming science and a set of tools for the ternary universe, which facilitate ease of use, efficient execution, and efficient maintenance. We need new programming models, languages, compilers and interpreters, dynamic libraries, as well as debuggers and optimizers. Parallel computing used to be applied only to high performance computers. Now it must be done to personal computers. Future computers may not have any single-core CPU chip; the hardware by default will be a parallel computer. Parallel programming is one of the most urgent problems in the information industry. Constructing efficient system software, including virtual machines, operating systems, programming environments and application benchmarks, for multicore and manycore architecture will be a difficult task.

The other important difference from traditional man-machine symbiosis caused by the ternary universe is the constant change of the application workloads and hardware platforms. To deal with these changes, many Internet service providers frequently upgrade their application software, as often as once a week. We do not yet know how to construct software and services that can adapt to changing environment, optimize, develop and evolve themselves automatically, according to changes in workloads and hardware platforms like a living thing.

The cyberworld is not only related to the time-domain and the space-domain, but also to the frequency-domain and the energy domain. Information thinking and informatization are not just digitization, but also involve control. Elegant algorithms and smart programming originate from comprehensively understanding of real problems.

**3. Difficulties in massive data utilization**

Although the exponential growth of many IT metrics will likely meet a technology wall by 2020, but not will data quantity. Currently, the whole world produce $10^{18}$ bytes of data annually. All kinds of data, including those of science and engineering, business, multimedia, Internet, are all rapidly increasing. All kinds of sensor networks and embedded systems will continue to produce even more data.

Storing, processing, transporting and displaying of mass data bring new challenges. Processing massive data offers a new opportunity and challenge for artificial intelligence. Some Internet contents providers have changed from the traditional "small-quantity-high-quality" way to a new one of obtaining good enough quality through massive data. Processing massive data with stochastic methods and high performance computing has become one of the new highlight of intelligent information processing. Services of transforming data to knowledge will likely become a pillar industry in the 21st century. However, there are still many fundamental problems that need to be solved to extract knowledge from data effectively, especially with massive, high-dimensional and imprecise data.

**4. Difficulties in realizing low-cost informatization**

In 2008, the Chinese IT market reached $112.1 billion and the number of Internet users surpassed 298 million. We estimate that 183 million of the users spent less than $300 annually, 157 million less than $150, and 70 million less than $16, all far lower than the level of developed countries. More than one billion people have no access to IT service and did not send even one email. The main reason is the high costs.

In the next 40 years, with the economy development and the increase of disposable income, over a billion Chinese people's propensity to consume IT will noticeably increase. But it requires us to reduce IT cost drastically to turn this propensity to demand. Low-cost terminals are the beginning of low-cost informatization. But, the total cost for users not only involves a cheap laptop. Other factors as listed below:

(1) Abundant and low-cost application services, to provide application value meeting user's economic and cultural demands.

(2) Ease-of-use, to reduce the learning curve and lower the barriers to use, maintenance, and innovation.

(3) Effectiveness, better cost-value ratio.

(4) Efficient utilization. Efficiency of the present IT systems, both in the server side and the client side, is very low. The average utilization ratio of servers is usually about 10 percent; even the utilization ratio of servers in many mega-corporations is only 7 percent.

The development of the historical low-cost technologies, such as PC, the Internet and the World Wide Web has shown that innovations in S&T are necessary for low-cost informatization. The following issues should be resolved:

(1) How to construct Internet software and services as assets, to enhance the reusability and the ability to seamlessly upgrade, so that software and services can improve their value/cost ratio in a way similar to Moore's law.

(2) How to create a harmonious and intelligent human-machine interface to improve the ease-of-use of the information networks.

(3) How to create a new architectures for the future information network systems to improve its effectiveness and efficiency. Today's virtualization work is only a beginning.

(4) How to construct an open information platform that provides users with freedom to use and to create information value.

A main difficulty in realizing low-cost informatization is to assure users' freedom of choice to avoid cost distortions caused by monopoly. The market domination of proprietary and heterogeneous computing platforms causes high cost, User's application developing requires the permission from the platform providers. The applications are bound to these platforms. Thus the users are locked in and lose the freedom of choice and innovation.

To realize the U-INS in China by 2050 to benefit the masses, we must create an open, standard basic platform, through innovation and a harmonious open industry-technology ecosystem. This is the common practice of important innovations in computing's history, such as PCs, the Internet, the World Wide Web, the Unix/Linux operation systems, the relational databases and the cluster architecture of high performance computers. In the recent 10 years, we see many successful Internet services such as searching, retail, auctions, travel, job-seeking and social network services. One critical success factor is that they all are based on the open and standard LAMP (Linux, Apache, MySQL, PHP) platform.

**5. Difficulties in reliability and security**

Although IT only penetrates 20 percent of the population in China and the informatization level in many industries is still quite low, reliability and security have already become a main concern of the society. The information network also brings with it new social phenomena and challenges. The tension is increasing between the demands for information openness and sharing and that for protection of personal privacy. The open nature of the Internet also leads to more spam and violence in cyberspace. As China gradually becomes an information society, the problem of how to assure the reliability of the information system has become more and more urgent. Reliability and security are related to not only the information system platform but also numerous application software and services.

We have neither enough scientific understanding about the reliability and security in the ternary universe nor enough knowledge about the relationship among reliability, security versus functionality, performance and cost of large scale information systems. We are short of effective engineering methodology to assure a system's reliability and security. This is a major reason for the much higher failure rates in large information network systems compared with electric power systems or building construction. We also lack good mechanism design to enable information network to become a social space of freedom for individuals and of harmony for the whole society.

**6. Difficulties in the construction of a universal basic platform**

Even if all the above-mentioned difficulties are overcome, we can not ignore another problem—dependence on foreign technologies. In the past 30 years of reform and opening up, great progress in the information market, information industry and IS&T have been made, but the dependence on foreign technologies has ever been very high. Main platform technologies, including microprocessor chips, operating systems, programming languages, databases, Internet and World Wide Web, were all invented in the past 40 years by international colleagues.

In the next 40 years, China's role as a follower and imitator should be changed. China should join the community of international innovators and construct platforms of four types: client platforms, server platforms, network platforms and sensor platforms. In particular, we should get rid of monopolies in these platforms, and create a solid foundation for network applications, especially data and knowledge services. To construct a U-INS basic platform with Chinese characteristics, we have to address the following issues:

**(1) Exploring future market demands.** The future IT market in China will be ten times more than the present one. We should not focus on legacy compatibility all the time, in particular, not be restrained by the monopoly platforms. Two persuasive precedents exist: high performance scientific computing and Internet services. Their orientation is to new demands and frontier problems with newly emerging users. They are based on open technologies, such as cluster hardware, Linux open source operating systems, Web technology and open source application software. They are not controlled by monopoly platforms. These two fields have been the most prosperous markets in the last ten years. In contrast, the desktop systems are highly affected by monopoly, and have made little progress in the same period.

**(2) Understanding the changing forms and contents of the basic platforms.** In the next 40 years, the platform for information networks will experience continuous changes. We should maintain the characteristics of openness, neutrality, multi-source of the basic platforms, and provide open source data, software, hardware and service of the basic platforms, in the key areas of user interfaces, programming interfaces, data formats, system call interfaces, instruction sets, network protocols, and key component devices.

**(3) Utilizing characteristic demands of Chinese economy and culture.** Future IS&T is both unified and diverse, like the natural world. We should study and understand historical characteristics (e.g., Chinese traditional culture) and contemporary characteristics (e.g., current Chinese economy characteristics) of the Chinese society and market. We should identify S&T problems, develop science and technology, construct a universal basic platform of information networks, all based on China's requirements.

**(4) Understanding and reducing complexity.** The U-INS is a complex system consisted of many parts. To realize low-cost informatization, we should positively utilize complex systems, but not allow construction complexity and usage complexity to spiral out of control. To understand and reduce the complexity will involve many challenging innovation tasks, such as: developing metrics of complexities that reflect the global characteristics; to develop network computing family and architecture by learning from the experience of computer family and architectures in decreasing the complexity of computers in the 1960s; to develop the theory of decentralized systems to enable a harmonious innovation environment, especially a loosely coupled ecosystem.

 

## Five Value Levels of Informatization

To measure informatization levels concisely, we can divide the information values into personalized value, expertise value, ubiquitous value, commodity value and basic value according to the information usage style and annual expenditure in IT per user. (*The following data on the annual expenditure in IT per user is based on the statistics in 2008*)

- **Basic value** (information poverty line, $150-annual for all hereinafter): It is the basic (lowest) information value that should be provided to everyone, including functions like personal productivity, surfing the Net, audio and video services, etc.
- **Commodity value** ($300–500): In addition to the basic value, people can enjoy some commodity computer products or net services. Here commodity means standard products and services targeted to a large user volume. A laptop or a blog service is an example to provide commodity value.
- **Ubiquitous value** ($500–1000): In addition to the commodity value, users can access the above services anywhere at anytime. Ubiquitous computing devices may bring new value, such as the location-based service brought about by mobile phones. Users can also enjoy services from sensor networks.
- **Expertise value** ($1000–10000): In addition to the ubiquitous value, users can enjoy value added information products and services as special members of a special expertise field. For example, a cartoon maker can get workstation hardware, animation software and deliver his or her services by cloud computing. In 2008, we estimated that the baseline expertise value in China was about $1000. As a comparison, in the United States the expertise value was $3000 in the medical field, $7000 for all private companies, and $15000 in the banking sector.
- **Personalized value** (>$10000): In addition to the expertise value, the user is given the centric position. IT can provide personalized services tailored to the user's individual needs. Currently, only a few large companies, researchers and wealthy individuals can afford this kind of value.

Of the 298 million Chinese IT users in 2008, we estimate that the percentages of users enjoying personalized value, expertise value, ubiquitous value, commodity value and basic value are 1.1%, 14%, 17%, 10% and 32%, respectively. 26% of the IT users are below the information poverty line. Furthermore, 80 percent of China's population are not IT users yet.

# 4 Building an Adequate Information Network

The major task of the development of the information technology in the next 10 to 40 years is to build an adequate information network. Through this network we can acquire information and knowledge conveniently, cooperate effectively and obtain a higher quality of life. Network technology undergoes broadband, mobility and tri-network convergence; it will move to next generation Internet and then to post-IP networks. Internet of things and sensor networks will evolve from integrated monitor-feedback-control to ubiquitous networks. Modern Internet-based service industry will rapidly develop and popularize. Network media will be a part of mainstream media, and the integration of traditional media and network media will build up a new structure to form medium public opinion. Cloud computing platforms will provide a supporting environment for different kind applications. Network science will develop from cross-discipline convergence to a well-formed network information theory. These technologies and services will influence each other, to realize an adequate information network.

## 4.1 Target and Roadmap

Short-term targets (2010–2020): Internet penetration will exceed 50 percent. On the basis of broadband, mobility and the tri-network convergence (integration of telecomm, Internet and broadcast), the next generation Internet should be developed in order to solve problems such as the shortage of IP addresses, security and manageability issues, and meeting the demands of new network applications. Modern Internet-based service industry will develop rapidly, and the integration of traditional media and network media will build up a new structure to form medium public opinion. The tri-network integration will evolve into "three screens linkage" (computer screen, television screen and mobile phone screen). The Internet of things and sensor networks will develop rapidly and enter our daily life. Breakthroughs in core techniques of post-IP network will build new architecture and protocols to overcome the IP bottlenecks.

Mid-term targets (2021–2035): Begin to deploy post-IP networks and get rid of the constraints of the TCP/IP protocols and architecture. Solve the problems of security, scalability and manageability more thoroughly. Satisfy the requirement of pervasive and ubiquitous networks, to form a trusted critical infrastructure. Web 3.0 semantic Web will see wide applications, and modern Internet-based service industry will be more user-friendly and convenient. People will be able to handle Web resources using natural language to obtain diverse personalized services. The massive application and use of Internet of things and sensor networks will realize the combination of land-sea-air-space networks and information integration. With adequate privacy protection, pervasive interaction space will be widespread, improving living and working environments.

Long-term targets (2036–2050): Post-IP networks will see mass adoption, to provide ubiquitous, sustainable, trusted and personalized services; to realize predictable and intelligent, ubiquitous sensor networks and embedded-body intelligent sensing, and to form intelligent ubiquitous information network systems (U-INS) for the masses. Virtual societies will become commonplace. 3D user interaction, entity interaction and brain-computer interaction will be widespread (as shown in Figure 4-1).

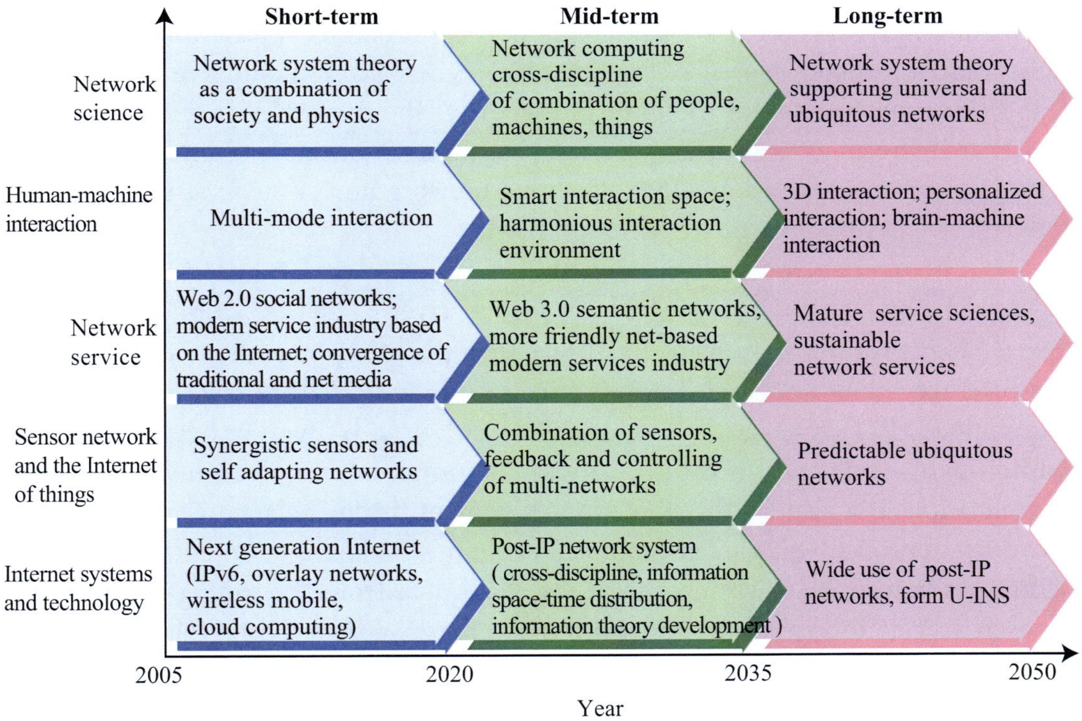


Figure 4-1 Roadmap for the information network development

## 4.2 Upgrading Communication and Information Networks

The Internet has inherent weaknesses in terms of security, scalability and manageability, and new network technology and infrastructure should be developed, whiling keeping the Internet tradition of openness. There are three approaches to do this. The first is evolutionary approach: evolve IP functionality via new RFCs. The second is the overlay approach, which is to build a new overlay network on top of the current IP network and permit significant flexibility in advanced service features. The third is the revolutionary approach: develop new "post-IP" networks. The evolutionary and overlay approaches are currently the main approaches used. It is likely that the international community will reach consensus on the new architecture and protocols of post-IP networks within the next 10–15 years. Much effort will be put into developing the revolutionary post-IP networks after 2020.

### 4.2.1 Upgrading Internet technology

Current limitations of the Internet architecture include security, robustness, manageability, quality of service, business model considerations, etc. Further evolution of the Internet is set to solve these problems while maintaining openness and autonomy of the Internet tradition. The target is to develop the next generation Internet before 2020 and promote the tri-network integration. IPv6 is the first step for the next evolution. IPv6 can not only solve the address problem but also improve on QoS certification and give the MIPv6 and IPsec mechanism for mobility and security. The current IPv4 is transferring to IPv6, to be an extensible and trusted ubiquitous Internet. In China, the CNGI project has constructed an IPv6 backbone network and some access networks; this plan will promote the transition to IPv6.

The overlay approach may include concept of an "IP knowledge plane" accessible by overlay. IP is pushed down to a "layer 3-" service, while overlay is "3+". This permits significant flexibility in support advanced service features. There are many successful applications such as P2P Internet phone, P2P download, P2P video broadcast and P2P VoD today. Overlay networks with intelligent nodes will not only provide a supporting environment for P2P applications but also for distributed management (traffic balance, QoS etc). The overlay approach will play an important role in developing the next generation Internet before 2020. To meet with the mass-data from the modern Internet-based service industry and network media, the cloud computing platform will be quickly developed and widely used. A large number of green and efficient data centers connected to the Internet will form the cloud. The cloud can not only provide the services of computing and memory to users but also provide a platform service (PaaS) and software service (SaaS). It is not necessary for users to buy all the software and hardware to get their own data center, platform and software like today; all these services can be obtained from the Internet.

The cloud computing platform can also provide support for the construction of overlay networks.

Before 2020, the industry effort will focus on the development of the next generation Internet, while universities and research institutes should actively explore the post-IP network system and find a new way that can be used to overcome the drawbacks of the IP protocol. We should start post-IP network research now to improve technology independence in China's future network systems.

In 2020–2050, the target is to realize new post-IP network systems and applications. The IP protocol is the basic protocol for the Internet; it has desirable characteristics such as structure simplicity and application independence from the underlying carrier network, which is the source for the vitality and innovations of the Internet. Yet, the IP protocol also has certain problems, including limited scalability and isolation, security and QoS, etc. As the Net advances, the technologies of underlying carrier networks and upper-layer applications have been significantly changed, and only the middle IP layer has not been changed (the so called Internet ossification), which has restricted the development of the Internet. The post-IP network has to break away from the restrictions of the current Internet protocol and system architecture. A new protocol will then be adopted in order to solve problems of security, scalability and isolation while also providing ubiquitous service to users. The problems that need to be addressed and key technologies that require breakthroughs include the following: improving the social trust of the Internet; overcoming the limitation of the current network system such as security, manageability, quality of service and scalability; adopting new technology such as sensors, mobile wireless, optics, etc., to meet the demands of various new applications in ubiquitous networks.

The current IP's shortcomings will not be resolved by conventional incremental and "backward-compatible" style designs. So, the post-IP network designs should be discussed based on a clean-slate approach. The design principle of the post-IP network can be set through reverse engineering, and get new ideas from cross-discipline research and innovation (there is striking similarity between the sandglass model of the Internet and the metabolism model in biology). We need to free from constraints by the current ICT technological systems, learn from social computing, economics, game theory, computational biology and quantum theory, consider time-space distribution of information, and extend current information theory to form new communication network frameworks.

The problems that need to be solved and key technologies that require breakthroughs include the following: new packet format that consider congestion and its control; new addressing modes; concurrency control and data distribution; support for mobility (dynamic addressing and location, location of mobile equipment and dynamic sources, etc.); new security mechanisms (protecting the routing structure against Denial-of-Service (DoS) and Byzantine attacks); new concepts of network management; new methods for tracking and fee-charging, the design of new routing: node with processing and storage capability, Non-routing functional equipment (security enforcement devices, encryption devices, application supporting devices); application-oriented cross-layer optimization algorithms and the model and rules for distributed application, etc. The research on the topology model for the new post-IP network system is another issue to be considered, for example, the tunnel mode besides the packet mode. The research test-bed should be built to support research on the post-IP network. The test-bed provides wired and wireless network environments to investigate the research results of new protocols and new architectures for post-IP networks.

### 4.2.2 The evolution of telecommunication technology

Before 2020, the development of telecommunication technology should be in accordance with the trend of tri-network integration and the convergence of communications with new media. Broadband, All-IP and FMC should be realized. Optical fibers will go on replacing copper wires gradually, and FTTH will become more popular. In mobile and broadband wireless communications, 3G cellular communications are evolving from HSPA (5(up)/14(down)Mbps) to LTE (50/100Mbps). The IMT-Advanced (4G) (100Mbps/1GMps) are developing. The popularization of IMT-Advanced will appear by the year 2020. The developed broadband wireless technology of IEEE 802.16e/m wireless metropolitan area networks will compete and complement each other with 3G/LTE, and they will be further combined into IMT-Advanced (4G). With the development of wireless communications, broadband wireless multimedia (BWM) in TV frequencies (UHF) should be developed to meet the demands of wireless city TV stations. To coincide with the demands of the post-IP network systems, all optical networks need to be developed after 2020. The key technologies of all optical networks are optical packet switch and optical routing. In broadband wireless communication systems, cognitive distributing and autonomous wireless systems should be further developed.

**Future Internet Design Initiative of the National Science Foundation (NSF)**

FIND (Future Internet Design) is a major new long-term initiative of the NSF NeTS research program. FIND invites the research community to consider what the requirements should be for a global network of 15 years from now, and how we could build such a network if we are not constrained by the current Internet—if we could design it from scratch. To solve this problem, FIND carried out wide research on network systems structure, design principles and operating mechanisms. FIND program encourages and fosters pursuit of rethinking the basic principles of Internet, such as:

- Whether or not to continue using packet switching mechanism?
- Whether or not to change the end-to-end argument?
- What is the underlying network architectural principles and architectural invariants?
- What is the underlying network architectural principles and architectural invariants?

## 4.3 Creating Ubiquitous Sensor Networks and Internet of Things

### 4.3.1 Strategic requirements of sensor networks

A sensor network usually consists of spatially distributed autonomous sensors which are typically equipped with communications device and microcontroller, base stations and control centers to monitor physical, industrial or environmental conditions. If a ubiquitous computer network can be regarded as the brain of the information world, the sensor network that connects the real world and the cyberspace would be the five sense organs and nervous system of the information world. The sensor network can realize efficient information transmission between the human and the machine as well as machine to machine. It is also the bridge between the cyberspace and the physical world.

The Cyber Physical System (CPS) is a network system that is used to drive the interaction between cyberspace and physical world. This system connects the cyberspace and the physical world with the sensor and performer and has the key ability of monitoring and controlling. The air traffic control system, avionics system, power grid, water carriage system and process controlling system are examples of CPS. These systems are also known as the Internet of things.

Researches on sensor networks and cyber physical systems have large overlaps. The former focuses on networks and sensors, while the latter emphasizes on computing and controlling; both are important components of future ubiquitous networks. CPS is a typical multidisciplinary area that concerns computers, semiconductors, networks, communications, optics, micro-mechanisms, chemistry, biology, spaceflight, medicine and agriculture, etc. The cross-discipline and interaction of sciences and technologies create the possibility for many new trends in the technology. For example, biology technology will depend on new equipment and the processing of biology information; the micro-engineering system, intelligent materials and new materials will make it possible for low-cost small sensors; the engineer will turn around to the biologist to understand the adopted methods to solve natural environment problems of the organism. These efforts can combine natural means and artificial devices to discover a better system than the existing one.

The progress of fundamental technology will further impel the progress of sensor networks and cyber physical systems.

The present network is composed of the personal computers and servers. The data stream comes from the center of the network to the edge. The development of sensor networks and cyber physical systems will move the data stream from the edge to the center of the network, and there will be more and more interaction and balance. There will not only need to be an exchange between files, but new exchanges, a new architecture and even new IT skills will be needed. The Internet of things will combine the machine and various things into the network.

As a typical application of sensor networks, Radio Frequency Identification (RFID) will be the main technology in future information storage, collection and processing. More and more international companies have taken part in this research. The attention placed on RFID development has been no less than any other emerging technology, and the development of this new technology will cause an increase in the quantity of server and terminal equipment.

The 3D sensor networks, which are constructed in roads, buildings, water areas, dangerous areas and interspaces, combined with distributing watch centers with powerful computing capability will forecast typhoons, heavy rains, floods, forest fires etc. For manmade emergencies, the networks will be good for quick decision making and disposition and thus decrease losses from natural disasters and manmade emergencies. Sensor networks can also be used to reduce energy consumption in intelligent homes, etc.

Sensor networks have evident advantages in information acquisition, cooperative sensors, high stability, small size, easy disposal, etc. They will be one of the supporting technologies for the realization of a safe, credible and intelligent road networks and urban rail transit systems. Sensor networks will be able to get real time information in a dynamic environment and cooperatively obtain information on the road. A comprehensive judgment can be made to predict and cope with traffic accidents and improve the utilization of the road. It can also be used to monitor emergencies and improve cooperative ability. The further developed "smart dust" can be embedded into any field in the world to get the combination of the information space and the physical world.

Sensor networks can be used in the military. Future soldiers will be dependent on a virtual battlefield system, which can monitor the position and condition of the soldiers as well as the weapons. This system can be used to lead the attack system to win precisely and complete the evaluation of the victory. This technology can also be embedded into the civilian arena; for example for child positioning systems, to get 24-hour care.

### 4.3.2 Scientific issues and key technologies of sensor networks

Sensor networks and Internet of things will meet with similar problems as those of other information systems. With the increase in scale and number of systems, the complexity of the heterogeneous network structure will be increased. As a new technology, it should be low-cost, green and energy saving, user-friendly and efficient. In the development of sensor networks, we have to face the problem about the coexistence of Boolean algebra and continuous algebra, which requires efficient uncertain inference, advanced complex system theory and heterogeneous network theory. It also gives us new requirements of the new sensing mechanism for the physical world. The key technology of sensor networks involves energy, intelligence, communications, standards, miniaturization and manufacturing, etc. The development of sensor networks and Internet of things requires the interaction of social sciences and natural sciences, including the integration of physics, chemistry, biology and cognition etc., and the synergy of science and engineering. It also brings forward a new challenge in terms of ethics.

## 4.4 Establishing Service Science and Providing Satisfactory Network Services

### 4.4.1 Strategic requirements of network services

If technology is the life of a manufacturing economy, then service is the soul of a knowledge economy. A new research field of service science has appeared. Service science is an interdisciplinary science that combines information science, economics, manufacturing and management, etc. Network information services are not only simple applications of network technology. They also relate to the exploration of science and key technologies in information services in the network environment. Web services are the most widely used network services; e-commerce, e-government, distance learning and telemedicine are all Web services in different styles.

### 4.4.2 Scientific issues and key technologies of network services

#### 1. Establishing a scientific system and engineering methodology for network services

The scientific system should be precisely defined and can be used to depict basic concepts (such as resource, data, information, knowledge, service, user and value), basic phenomena (laws), and computing models (such as a service machine model beyond Turing machine). It can also be used to analyze and synthesize complex systems of network services. The essentials of network service are network service based on knowledge, which requires the combination of explicit and implicit knowledge and needs to resolve the contradiction between information asymmetry and the demand of information transparency. We need to develop Web science methods and use them in the digital community, digital government and digital society. Allow a virtual-and-realistic society to be the common social norm and an effective mechanism for the establishment and realization of a harmonious society.

**2. Realizing personalized network services**

To construct the means and methods for analysis of the behavior and mentality of users in the Web space and the population dynamic model in order to analyze the existing form of the network and to predict the probable form in the future. To construct universal and domain-specific knowledge ontology systems and knowledge grids, and form standards and protocols, in order to enhance the efficiency of traditional fields and the standard of living for all people, promote a personalized modern Internet-based service industry and new media. Establish methods for evaluating and improving service value. Effectively manage the network space and ensure legal rights inside or outside the cyberspace.

**3. Constructing efficient and reliable network service execution systems**

To invent new technology to reconcile the computer systems flat methodology, the network system's protocol stack methodology and the users' interaction methodology, and ensure convenient dynamic service programming, service certification and service composition. To develop a scientific system and engineering methodology for a sustainable network service restricting increases in energy costs, resource costs and emission. Develop a mechanism for social management and safeguard for the network service. To allow the network service to encourage the user to participate in value creation through mechanism design. This will not only enhance service quality and efficiency, but also provide safeguards for personal privacy.

## 4.5 Developing the New Information Theory of Network Science

### 4.5.1 Requirements of network science

Network science is an emerging science that is used to study the principles, models and computing theories in the ternary universe of the cyber space, the human society and the physical world. It involves interdisciplinary science problems in the economy, society and information science. When the basic principles, models, mechanisms and algorithms are formed, network science will complete the process of integration and form a relative steady discipline.

### 4.5.2 Fundamental issues and technologies of network science

The current methodology of network science research mostly draws from statistical physics. With the development and perfection of network science, principles and mechanisms of quantum computing, game theory, biological computing, social computing and other related areas will probably be emerged into network science. The development of network science will be coming out with the development of technologies about information physics, bionic network and social network. Those paths will combine to further research and form a basic information theory for network science.

Generally, the basic aspects of network science are related to the theory of network information, network system and network computing.

**1. Information theory of network science**

Shannon's information theory seems ineffective in the face of complex network phenomena. First, finding common and intrinsic characteristics and mechanisms of the complex networks is critical for network science. It is predicted that some new common rules and mechanisms will be uncovered in the next decade. These findings may be composed of theoretical system of network science. These would be expected to uncover the relationship between the micro-scope features and the macro-scope features of complex networks. Economical, social and physical mechanisms need to be addressed after the basic rules and mechanisms of network science reach to a common agreement. The research will likely be carried out according to the following principles: Dynamics, finding the reason for network emergence and evolving phenomena; Theory of control, finding the inherent rules in the macro-scale of networks; Theory of new general network security: a theory about the effect of network stimulation (perturbation) on the development of networks.

**2. System theory of network science**

The development of network science is driven by the human society, the physical world and the cyber world and it will make the information network more intelligent, effective, adaptive and self-healing than it is. Another key issue for network science is the systematic theory of the network. The system theory of network science may be composed of the shape of networks, value theory of network systems, topological theory of the network system, and method of mathematical physics in network systems.

Computing theory of network science: Network computing theory involves intelligent computing, structural calculation, content understanding, perception computing, etc. Network computing theory needs a new computing model that regards the whole network as a subject. The locality theory is potentially fundamental to the network computing theory. Quantum computing will likely become an active branch of network science. It is predicted that the research branches of game theory, emergence analysis, network dynamics such as diffusion and synchronization, parallelized optimization of network algorithms will be popular topics within 20 years.

## 4.6 Achieving Natural and Pervasive Human-Computer Interaction

### 4.6.1 Requirements of human-computer interaction

Human-computer interaction (HCI) is one of the key fields in computer science and engineering. In some sense, the goal of our research on computers can be divided into two categories: one aims to improve performance while the other targets to make it easier to use. The purpose of human-computer interaction is to enhance the user's experience, which is composed of the following three aspects: manipulation of the equipment, management of the information system and usability. After the appearance of computers, HCI migrates from punch card (tape) and character terminals to WIMP (Window, Icon, Menu, Pointer). Pervasive interaction will be a new kind of interactive environment after browsers/Web. Pervasive interaction emphasize a human-centered approach and benefits from cloud computing, wireless communication and sensor networks. In a pervasive interactive environment, the equipments should be easily connected to each other, achieving sharing data, providing active services, invisible computing resources, equipping natural interactive interfaces.

One of the trends in HCI is to achieve more natural multimodal interaction. In the coming decades, multimodal HCI will be widely used from desktops, laptops to handheld devices. With the popularization of ubiquitous computing environment, active sensing to both user and environment and smart interaction will become important features of smart spaces, and invisible computing needs ubiquitous interaction. Tangible interactive interface, designed for special purposes, will play an important role in design, command and entertainment systems. The brain-computer interface (BCI) will be utilized and popular in wearable computing systems. As an important engine to push the progress of computing technology, HCI is one of the key players in computer science and technology. R&D in HCI directly promote IT industry and the effect of the application. In the past decades, China paid little attention to research of HCI, and as a result, there is a big gap on business model, design, and applications between China and the IT leading countries. With the progress of the China economy, new and special requirements will be put forward on human-machine interaction.

### 4.6.2 Prospects of human-computer interaction

**1. Future display technology**

Information display is one of the fundamental IT technologies, and as important as information processing, storage and communication. It has a huge market of more than a hundred of billions dollars. Vision provides us more than 83% information that we capture from outside. Therefore, display is a fundamental HCI modality. Display now has more breakthrough opportunities than IC technology. Some competitive display techniques are as the following.

(1) Organic Light-Emitting Devices (OLED). Comparing with the LCD, OLED has some advantages: light-emitting without backlight, higher contrast, wider view angle and lower delay, etc. It can be made into large, thin sheet. These features make OLED a promising next generation flat display technology.

(2) 3D display technique. In the past years, people can only immerse in the 3D world with 3D glasses. However, some displayer providers deliver glasses-free 3D display devices in recent years. This means that the 3D display screen will be popular on TVs, computers, and even mobile phones in about 10 to 20 years, and the users can enjoy 3D virtual scene without 3D eyewear.

(3) Electrophoretic displays (electronic paper). Electrophoretic technology combines the advantages of traditional paper and electronic displays. It consumes much less power than other flat displays. The manufacturing process is relatively simple than other competitors. It can be in the form of flexible displays. The electronic paper can be produced onto the substrates of plastic, metal, and glass. The electronic paper will replace the LCD displays in ebook reader, and will bring a revolution in book reading.

(4) Flexible displays. The folding-able or coiling-able displays bring those portable terminals an opportunity to equip a large screen, and provide more chance to fit fancy industry designs.

(5) Projector based I/O. The keyboard of the terminal can be projected onto a panel and the user can input on the panel. In the future, the projector can be integrated with laptops and mobile phones.

(6) Laser display. Currently, the color space in off-the-shelf displays can only cover about 83% of range that our human being's vision can percept. Most of the displays can not completely present the original natural color. This became a major obstacle for the further improvement of display quality. However, the source of high saturated color of red, green and blue lasers solve the problem, and can represent a large color space. The lasers can cover about 90% of range that the human being's vision can percept, and it can represent natural color perfectly. Potentially, the laser display will become the fourth display technique after black and white display, color display and digital display.

**2. Multimodal interaction**

With the progress on speech recognition, image understanding, and multi-touch, the means of traditional interaction will be greatly enhanced. The traditional HCI which is featured with keyboard and windows etc. will keep dominating in the field of the computer as a working platform, however, the intelligent technologies will improve efficiency of interaction. With the improvement of the semantic gap and dynamic updating corpus from web, the input efficiency of the keyboard will be greatly increased.

The further development of digital ink will allow the extension of the pen with force feedback and enable a real writing experience in signature, handwriting, and even painting. Computers and traditional art will be integrated more tightly.

In the fields of education and entertainment etc, multi-touch technology will be popular in ten years. By about 2020, multimodal interaction characterized with the gesture command/control and voice interaction, etc. will dominant the HCI in the above applications. Meanwhile, interaction in 3D space will become a kind of HCI modality. Emotion understanding will become an important perception way to understand user's status and to provide personalized interaction. In addition to the traditional keyboard and the mouse, multiple cameras and microphone array will become standard configuration of a computer before 2020, and these will supply the basic hardware for multimodal interaction. Meanwhile, the personal high performance computers will provide enough computing resource for multimodal interaction.

As to the output, stereo displays, augmented reality which merge physical world with virtual world, and large touch screen displays will meet various purposes, such as design, education, entertainment, etc. Glass-free 3D display devices will be popular before 2020. Meanwhile, the volume display will be in the time to market.

**3. Ubiquitous interactive environment**

Pervasive or ubiquitous computing will provide a broader information space. The computing, information resources and multimodal sensors will be embedded/embody in the smart space. This human-centered working and living space will equip the natural and human-like interactive interfaces, which provide the user with the service easily from the invisible computing resources. Working / living in such a smart spaces is to enjoy active service from the ubiquitous computing environment. In this process, the computer is not a simple tool which executes the instruction from the user, but rather a helper that can actively provide personalized service. Smart space will be gradually merged into our house, conference room, and offices within the next 40 years.

A perfect pervasive computing environment can identify who is the user. It can percept user's emotion, understand user's intention in some sense, and provide right services. Users can conveniently interact with all kinds of equipment and get the required information. User's wireless devices can connect to the infrastructure seamlessly. This enable user to acquire information uninterruptedly, and record daily experience in the space. The record can be retrieved thereafter. The pervasive environment supports multiple users to cooperate in the same virtual space. This kind of interaction space can be smart meeting room, smart house, or smart classroom, etc.

It is predicted that smart meeting rooms will be the first prototype of pervasive interactive environment in the coming ten years, and the smart rooms will be able to offer the functions including automatic recoding, statistics analysis, sorting record in various dimension, perception to the participants, providing discussion space for small groups. From the view of equipments, the interactive desktop and interactive touch wall will be widely used. Meanwhile, some city scope interactive functions, such as location based services, will be gradually merged into our daily life.

The smart space with pervasive sensing and interaction will merge into family's daily life around 2020. The house with smart functions such as automatic access control, appliance control with speech and gesture will greatly improve living quality. It is predicted that the pervasive smart space will be available in city-level around 2030 when the user's privacy can be well protected,

**4. Interaction for special need**

For those who suffer from brain paralysis or the user of the wearable computers, they can't either use the traditional keyboard, speech and gesture or express their intentions easily, for example in some games. Therefore, brain-computer interaction (BCI) and tangible interaction will be rising stars in special needs.

The goal of BCI is to obtain the electronic signal from the user's brain with EEG or other equipments and control various devices. The key issues in BCI include the following:

(1) Acquiring robust brain signal easily. Even though equipment such as EEGs, fMRI, etc. can be used to acquire the signal, they are still far from real world requirement. It is predicated that head-mounted EEG for BCI will be used around 2020, and BCI controlled devices will be on the market then.

(2) Adapting of brain signal across different users. Comparing with other recognition problems, because the brain signals across users are quite different, how to generalizing the BCI to different user is a big challenge to BCI.

Tangible interaction can provide interactive means by directly using the tangible thing as an object. With popularize of wireless sensor, tangible interaction will be widely utilized in games and design, etc. The combination of tangible interaction and 3D display technology will enable the construction of an augmented interactive environment.

### 4.6.3 Open problems and key technologies of HCI

(1) Semantic gaps. Multimodal interaction is different from previous interaction with keyboard and mouse. Filling up the semantic gap will be the key issue to minimize the uncertainty of interaction.

(2) Understanding of the single modality interaction and integrating of multimodality interactions. Single modality such as speech, gesture can provide some means of interaction. However, in our daily communication, different modalities are concurrent, and they are complementary. For example, the communication during the conference, interaction accompanies with speech, mood, tone, emotion, hand gesture, and even body language. It's shown that multimodal information can greatly improve efficiency of interaction.

(3) Challenges to computational performance. In the past 10 years, HCI has consumed more than 60% of the increasing ability of CPU power for those functions such as online grammar, lexical checking and online dictionaries, etc. Multi-mode interaction requires large scale data processing which need to consume computational power. It can be predicted that multi-mode processing

will take up most of the hardware in the future personal high performance computer. To decrease costs caused by HCI and make multimodal HCI easy to use, it's necessary to develop ASIC and new kinds of terminals.

(4) Human in the loop. Humans are the dominant factor during interaction. The quality of interactions could be improved by adopting the user's feedback.

(5) Enhancing interaction with the Web and wireless network. The performance of interaction can be improved by understanding the interactions with information from web. Some techniques can help to improve HCI and need to pay more attention in near future. These include resource discover and connection, localization from multi kinds of signals, context-aware, application/ cost aware wireless connection.

(6) Remote collaboration. The other function of HCI is to connect people in different location and collaboration based on HCI. The collaborative response of users in the collaborative working space is the key for remote cooperation. To achieve efficient remote collaboration, an open collaboration platform which allows users join easily is necessary.

(7) User experience. Keeping improve the user's experience is one of the key issues of HCI. Comparing the signal modality HCI user study, larger testing spaces and more samples should be supplied for user experience study in multimodal environment.

(8) Interaction security. Because of the information exchanging in the process of interaction, interaction has been the weakest security link after the Internet. The input and output of password and the user's information will lead to problems in safety and privacy protection, especially when interaction is not only run from the traditional wired-connected equipment such as the keyboard, the mouse and the display but also from wireless-connected ones. With wireless equipment, security becomes a more serious problem. Interaction security will be a hot topic in information security in next a few years. Meanwhile, dynamic authentication, encryption will be widely adopted in HCI devices.

# 5 Revolutionary Upgrade of Information Devices and Systems

Traditional information devices and equipment systems are encountering enormous obstacles in complexity, cost and power consumption. In addition, CMOS-based chip technology is approaching its physical limit, and the industry is eagerly looking forward to disruptive new technologies. On the other hand, future chips will be integrated with various functions such as computing, storage and communications, and characteristics such as more varieties and shorter design cycle, all of which need to search for a new approach. From 2010 to 2050, the application demand for information devices and equipment will experience substantial growth, even though it has not yet emerged as the same dominant technical route on par with CMOS integrated circuits over the past 30 years. Related scientific and technological developments have reflected the uncertainty and diversity in the industry today.

In addition to the continuous upgrade of current electronic computing technology based on CMOS chips, other forward-looking computing technology research, such as quantum computing, spin computing, molecular computing, DNA computing and optical computing, is being vigorously carried out.

As frequency increase is limited by power consumption and other factors, academic and industrial circles are paying increasing attention to parallel processing from all levels. Parallel processing and intelligence are two core technologies in most need of breakthroughs in information system design. Over the past several decades great efforts have been made in these areas, but only limited progress has been achieved. The information society is composed of small and large information systems, and system architecture is both an old and a new field. Increasingly complex systems, growing diverse demands, ever-smaller space requirements, increasingly lower costs and ever-smaller power consumption have all led to major changes in the science and the technology for building information systems.

## 5.1 Targets and Roadmap

Short-term targets (2010–2020): in micro-nano electronic technology there are three kinds of development approaches including More Moore, More than Moore characterized by functional diversification, and Beyond CMOS targeting devices of a revolutionary nature. The development of optoelectronic and photonic device technology integrated with nano-electronic technology will give birth to such technologies as processor array chips supported by on-chip interconnections, digital-analog hybrid optical system integrated in chip packages, and integrated optical computing chips. The study of the fault-tolerant quantum computer structure can achieve a quantum simulation that classical computing cannot simulate. The development of data centers and super-servers characterized by low costs and low power consumption can support the deepening of information network application and reach Exaflops level ($10^{18}$ flops).

Medium-term targets (2021–2035): define the new mainstream device technology that can substitute CMOS. Succeed in researching small and medium-sized quantum computers and carry out the real application research of quantum computers and new quantum algorithms. Invent new computer systems architecture and software, overcome the challenges of massive parallelism and reliability, effectively support data and knowledge services and achieve the goal of sustainable computer systems with zero growth in energy consumption and emissions.

Long-term targets (2036–2050): revolutionary devices are widely available. Overcome the problem of "scalability" in quantum computing systems, succeed in producing "general purpose" quantum computers and provide practical applications. Research on super-servers that can adapt to changes, and self-optimization. Achieve the goal of supercomputing at Zettaflops ($10^{21}$–$10^{24}$ flops) level. Satisfy the personalized needs of mass users (as shown in Figure 5-1).

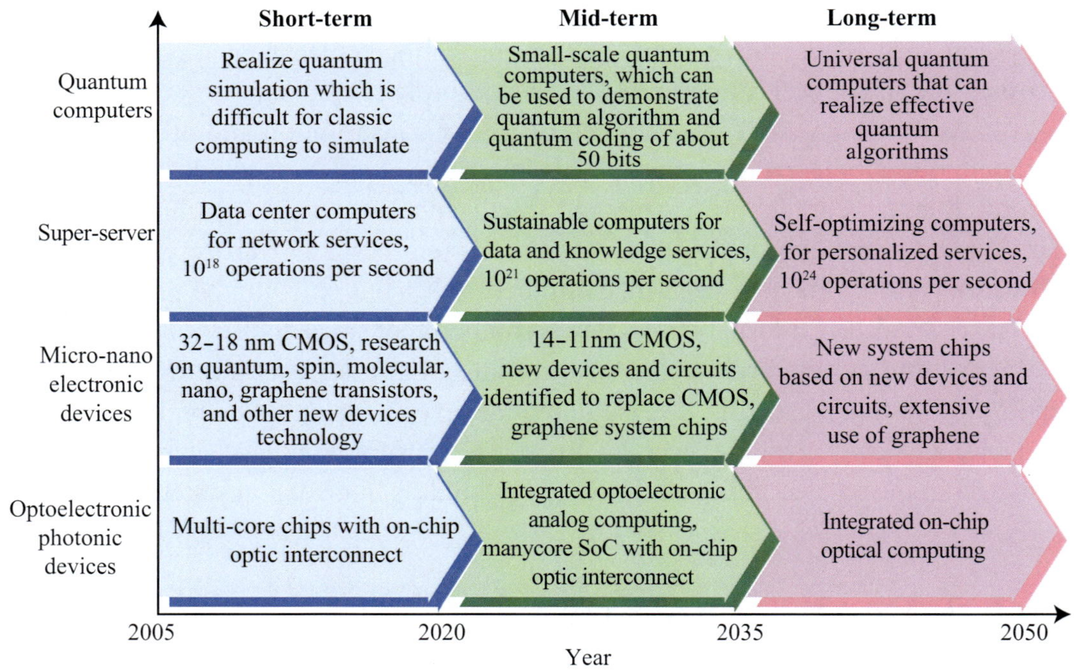


Figure 5-1 The overall development trends of devices and system upgrades

## 5.2 Three Paths for the Development of Micro-Nano Electronics

Since 1965, silicon microelectronic technology has been guided by Moore's Law and the principle of scaled reduction, which has enabled IC integration and performance to double every 18 months. Under the guidance of Moore's Law, the IC industry has experienced rapid development, allowing electronic devices to become faster, smaller, lower in price and more complex. Great progress has been made in terms of transistor sizes from micron to nano-scale. The main component of the transistor - the gate dielectric - is expected to achieve the scale of several atoms when its physical dimension has reached 11nm or smaller. Silicon CMOS technology will be subjected to such restrictions as its basic physical properties, process technology, huge investment needed and other factors related to speed, power consumption, integration, cost, processing uniformity and reliability, etc. Therefore, one of the most significant scientific and technological problems in the $21^{st}$ century will be in performing continuous innovation in such basic research fields as nano-scale device physics and new materials, and to seek breakthrough solutions in ultra-fast speed, ultra-low power consumption, sub-11nm basic devices and integration technology.

The development of micro-nano electronics in the Post-Moore era will witness three major trends:

(1) More Moore: continue to scale down the physical size of CMOS devices and increase integration, improve the circuit performance through the application of new materials and device structure innovation, and simultaneously continue to promote system-on-chip (SOC) technology based on information processing digital circuits.

(2) More than Moore: in the past, the reduction of physical size and the increase of the device density was constantly pursued, but now more attention is being paid to increasing the function diversity of system integration by means of diversified devices represented by system-in-package (SiP) and function integration. The end result will be revolutionary changes in mega-function electronics.

(3) Beyond CMOS: traditional silicon-based CMOS technology will reach its performance limits in around 2022. Explore new principles, new materials and new structures, and direct development towards nanometer, sub-nanometer, as well as multifunctional devices, such as carbon-based nano-devices, quantum, spin electronics, and molecular devices.

### 5.2.1 More Moore

Moore's Law predicts that, before 2021, the development trends of IC technology will be as shown in Table 5-1:

Table 5-1 Development trends of CMOS integrated circuit technology (ITRS 2008)

| Year | 2012 | 2013 | 2014 | 2015 | 2016 | 2017 | 2018 | 2019 | 2020 | 2021 |
|---|---|---|---|---|---|---|---|---|---|---|
| Feature size (nm) | 40 | 35 | 32 | 28 | 25 | 22 | 20 | 18 | 16 | 14 |
| Wafer size (mm) | 450 | 450 | 450 | 450 | 450 | 450 | 450 | 450 | 450 | 450 |

To continue developing CMOS devices, the introduction of new materials has become more and more popular. Such innovations as silicon oxide replaced by high-permittivity dielectrics, polysilicon gate replaced by metal gate and silicon channel replaced by SiGe strained silicon channel materials have shown their advantages in reducing leakage and improving speed. Currently, new high-mobility channel materials are being researched such as silicon-based germanium or III-V family semiconductor materials. In recent years, the emergence of the carbon nano-material Graphene, an alternative to silicon, has been considered to be the hope for More Moore. This is because Graphene not only features such superior semiconductor electrical properties as high carrier mobility rates and adjustable band gaps as well as high thermal conductivity, but more importantly its thin-film configuration is compatible with current silicon plane technology, thus enabling large-scale integration.

Key scientific and technical issues:

(1) Scientific and technical issues of conventional CMOS devices include the breakthrough of new nano-scale micro-processing technology; high-permittivity insulating layers, metal gates, strain channels and SOI liner materials; new technology and new structure leakage-reduction devices; research of CMOS devices from new materials, including the growth of silicon-based high-speed channel materials; research of heterojunction materials and interfaces.

(2) Research non-planar CMOS devices (such as double-gate FinFET and tri-gate transistors); leave no stone unturned to find ways to continuously decrease the size of CMOS devices.

(3) Carry out research on large-area growth of two-dimensional graphene crystal thin film and epitaxial selection, device structure and process technology, as well as the effective control of the basic physical model computing and electrical properties of materials and devices, while at the same time carrying out basic research on other new film substrate CMOS devices.

(4) Strengthen innovations with regard to system-on-chip architecture design (such as low-power circuits and multi-core MPUs).

(5) Research multi-core-based chip technology as well as the design of NOC (Network on Chip) with CPU/DSP core as an Internet node.

### 5.2.2 More than Moore

In 2007, ITRS formally listed the functional diversity represented by SiP as the new direction for the development of semiconductor technology. The future system-level package will be a complex package integrated with digital signal processing, analog/RF, passive components, high pressure, high power, sensor/actuator, optoelectronic devices, bio-chips, sensors and micro-energy, etc. SiP features such advantages as more new features, good compatibility with a variety of technologies, strong flexibility and adaptability, lower cost and a shorter development cycle. SiP will advance from traditional two-dimensional integration to three-dimensional integration.

Because of the complexity of SiP, higher requirements have been put forward both in terms of design and process technology, resulting in great changes to traditional post-packaging technology.

Key scientific and technical problems:

(1) The design of such components as IP cores, IC chips, power and passive components as well as layout with the aid of computers; the design of the systems testability; the elimination of oscillation, overshooting and crosstalk, and radiation, as well as the consideration of heat dissipation and reliability in the design of high-density wiring; substrate material optimization and selection, etc.

(2) Carry out research on the revolutionary process technology for system-level packaging. The advanced wafer-level package adopting such front-wafer process technologies as lithography produced by IC wafer and metal-based transformation, the flip-chip bump welding with extremely narrow pitches, the new interconnection technology based on TSV, the passive and active device embedding technology achieved by thin-film interconnection, the three-dimensional chip-stacking technology and other three-dimensional integration technologies.

### 5.2.3 Beyond CMOS

In accordance with the trend forecasting by the International Semiconductor Industry Association, the size limit of CMOS is roughly 10nm, with another 15 to 20 years needed to reach this limit. Electronic devices in the information arena will face essential technology updates; hence it is necessary to research a new generation of basic devices. The main research areas for revolutionary devices focused on by the forefront of world science frontier include the research of nano-tubes, nano-wires, nano-devices and circuits; the research of quantum flow, quantum dots, quantum-bit devices and circuits; the research of spin materials, devices and circuits; the research of molecular materials and devices; the research of direct self-assembly materials, devices and technology; and the exploration of other new principle devices with application potential.

The main research focus:

(1) When device sizes are close to the wavelength of electron, the effects of wave can no longer be ignored, and nanometer-scale electrons begin to show some quantum effects such as quantum tunneling, quantum interference, and the Coulomb Blockade Effect. The design and preparation of different types of quantum devices based on the above quantum effects (such as single-electron transistors, quantum cellular self-adaptive devices QCD, quantum memory, and quantum-bit quantum computers) is one of the key directions for research.

(2) Electrons possess the two properties of charge and spin. Conventional electronics adopts the electron flow generated by electric charge motion due to electric fields, with the charge interaction energy at eV level; spintronics adopts the electron flow generated by changes in electron spin directions due to the magnetic field, with the spin interaction energy at meV level; hence extremely high speed can be achieved at very low power consumption. Further study of such scientific issues as spin injection, spin detection and spin transport indicate that spin devices, which is the integration of ferroelectric materials and semiconductors, will be the direction for future research. The freedom degree of electron spin can be used as an information carrier, or quantum bits, which can be applied to quantum information and quantum computing.

(3) Molecular electronic devices consisting of organic semiconductor molecules can control the electronic motion under the effect of a specific working mechanism and then achieve the devices and circuits of special functions. Disciplines of molecular electronics include research on the working mechanisms of molecular devices, the design and synthesis of functional elements, crystal growth, the preparation of ordered thin films, the structures and properties, special physical and chemical phenomena and processes, the assembly of molecular devices and other related scientific issues.

(4) In addition to significant quantum effects, special attention is being paid to the unique electric, magnetic, and metallic and semiconductor characteristics as well as a wide range of prospects for application that nano-tubes, nano-wires and nano-belts themselves possess. The research focuses on the growth and self-assembly technology of nano-tubes, nano-wires and nano-belts, and the structural design and preparation of transistors, memory and logic devices.

The above new-principle devices can not be separated from the exploration of new materials (such as ferroelectric and complex metal compounds, molecular and strongly correlated electronic states and other materials), hence the physical, structural and self-assembly preparation of new materials is a common subject for new device research. It is difficult to establish a clear roadmap for development for currently emerging beyond CMOS technology. We should accelerate research of nano-electronic device schemes and define materials and devices that are possible replacements for CMOS as soon as possible in order to concentrate on achieving an early breakthrough.

### Graphene Electronic Device

IBM announced that they have developed the graphene transistor with gate length of 150nm and cut-off frequency of 26GHz in January 2009. The encouraging graphene transistor may drive Moore's Law to continue, and also become the new hopeful research area beyond silicon-based CMOS. Graphene is a one-atom-thick planar sheet of carbon atoms that are densely packed in a honeycomb lattice. Graphene is of high carrier mobility. The electron and electron hole have the similar carrier mobility which is 106cm$^2$/Vs in theory. The experimental result is as much as 15 000 cm$^2$/Vs which are 10 times of the electron mobility and 30 times of the electron hole mobility of Si.

It is expected that graphene material and transistor of high performance is able to be developed in around 2020, and the interconnection and integration issues can also be resolved. Graphene SoC is able to be developed and commercial produced in around 2035. This will enable carbon-based CMOS to replace silicon-based CMOS that has dominated IC for more than 50 years and bring about revolution on IC.

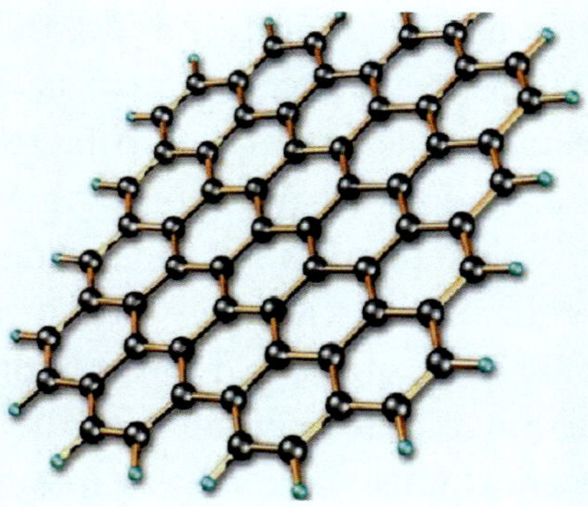

Source: http://www-03.ibm.com/press/us/en/pressrelease/26302.wss

## 5.3 Developing Revolutionary Optoelectronic and Photonic Devices

### 5.3.1 The strategic requirements of optoelectronic and photonic devices

Traditional optoelectronic and photonic device technology has consistently been one of the basic technologies supporting the rapid development of optical network communication. With future optical network communication technology directing the development of high speed, large capacity, multi-service integration, intelligence, personalization and ubiquity, it will continue to provide hyperfine wavelength division multiplexing (Hyperfine WDM) technology and achievements in optical network's security, reliability, controllability and manageability with the best solutions on the component level.

Revolutionary optoelectronic and photonic device technology has emerged with today's demands with the integration of computing technology, optoelectronic technology and optical computing technology. In order to satisfy the requirement of explosively growing computing and information processing speed, in addition to CMOS-based integrated chip technology still advancing towards the continuous scale down of physical size in accordance with Moore's Law, people are increasingly hoping that future chips are "full of light", and integrate today's silicon processing capabilities with the unmatched speed of light. Light can not only travel faster than any other substance, but there is no such interaction between photons as that between electrons and data can pass directly through between several independent beams, therefore incredibly large quantities of data can be transmitted through the same narrow transmission channels.

Based on the above expectations, future chips will quite possibly be integrated with such functions as computing, storage, communications and information processing. This demand has driven the integration of CMOS-based electronic technology and photonic technology. The continuation of Moore's Law in the future will be mainly subject to the limitations of huge industrial investment risks, because it will cost as high as several or even ten billion dollars to establish a modern chip factory. Facing such a large industrial investment risk, simply excavating the potential of chip technology bit by bit will no longer satisfy people's needs, forcing people to make a fresh start and seek new revolutionary changes. The integration of computing technology, optoelectronic technology and optical computing technology not only guarantees the sustainable development of the CMOS integrated chip industry and avoids the enormous waste of resources and industrial risks that would result from replacements and upgrades, it can also allow existing CPU and DSP to increase in speed by over 1000–50000 times. It can also promote revolutionary changes in computing technology which has become the new generation of chip technology, with the integration of computing, storage, communications and information processing. In the end, it will promote the realization of on-chip optical computing technology.

### 5.3.2 Scientific issues and key technologies of optoelectronic and photonic devices

Comprehensive viewing of the development of optoelectronic and photonic technology, besides the traditional technology supporting the rapid development of optical network communications, an even more important trend in the future is the integration of CMOS technology, optoelectronic technology and photonic technology, which, both in multi-core processors MP adopting SIMD instruction system structure and in MPP chips adopting massively parallel processing element array, or in film optical coating computing and in optoelectronic DSP chips achieved by using optical vector-matrix simulation computing, will show superior data processing capabilities, thus becoming a far-reaching revolutionary technology. Related scientific issues include the physical mechanism of ultra high-speed electro-optical and optoelectronic conversion, the sub-wavelength transmission properties of surface plasmon polariton (SPP) and the physical mechanism of load entanglement transmission, the optical nonlinear effect based on SPP in weak light, the quantum manipulation of light and spin in solid, and the new physical mechanism and meta-materials for the realization of spatial light modulation. The following three key technologies are worthy of note: processor array chips supported by on-chip optical Internet, the development from optical simulation computing to packaged optical systems, and on-chip-integrated optical computing technology.

**1. Processor array chips supported by on-chip optical Internet**

The advancement of optoelectronic and photonic technologies will drive the CMOS system on Chips (SoC) development into on-chip optical Internet-supported processor array chips. The SIMD (single instruction and multiple data architecture) computer proposed in the 1980s is actually a kind of processor array machine interconnected via a simple network topology. The constant innovation of silicon integrated technology makes it economically viable to design SIMD computers on single chips with hundreds of processing elements and embedded memories, and to a certain extent, also solves the difficult problem of whether instructions can be accurately transmitted to thousands of processing elements within each clock cycle. As computer architecture design has transferred to chip design, whether they are multi-processor system on chips (MP SoC) with processors as IP cores, or massively parallel processing system on chips (MPP SoC) achieved by means of processor arrays, they all need to achieve high-speed data switching and routing between different IP cores. This kind of MPP SoC, which can simultaneously exchange data via reconstructing Internet, is actually a prototype of the on-chip Internet.

So far, reconstructing Internet SoC chips are achieved by means of copper wires. However, due to the large power consumption and long time delays of copper wire interconnection, the array dimensions of MPP SoC will be limited. This also limits the capability of increasing the computational speed by means of large-scale parallel array. Therefore, it is quite natural to put forward the concept of whether or not the optical Internet can be used on chips. It is crucial whether or not the ultra-high-speed electro-optic conversion technology can be developed, and if not, the ultra-high speed characteristics of optical interconnections cannot be realized. Currently the industry has achieved a 40Gb/s electro-optical conversion rate by means of a single wavelength, and if the optical transmission bandwidth can be fully utilized, a 1Tb/s electro-optical conversion rate can be achieved. This kind of electro-optical conversion coupled with optical interconnections can be transmitted 100 times faster than copper interconnections, and it consumes ten times less power.

Key technologies include: a reconfigurable massively parallel system architecture; a data distribution and configuration program; design of array processors, memories and array routers; optical interconnection and its integrated chip technology; ultra-high-speed electro-optic data conversion technology; electro-optic, optoelectronic component and its array technology; flip chip bond process technology and so on.

**2. The development from optical simulation computing to packaged optical systems**

Optoelectronic and photonic technology will enable optical simulation computing to develop from the module level to the optical system in package (OSiP). Real-time image processing, video compression and the processing of radar and remote sensing system signals will require a great deal of parallel multiplication of vectors or matrixes, which will also bring difficulty in the serial operation of massive binary addition. If optical vector-matrix multiplication is adopted, the workload of signal processing will be greatly simplified. A traditional and optical component-based system can in part achieve the above signal processing work, but the versatility of such a system is reduced because it doesn't have programmable capabilities. The semiconductor vertical-cavity surface-emitting launcher (VCSEL) line array with an adjustable brightness level is adopted as an input vector matrix, all necessary computing are completed by an open space light modulator (OSLM) of the two-dimensional array set by programming, and finally it is up to the light detector array to detect the results of the vector-matrix operation, which when applied together can complete programmable optical simulation computing.

Low-light-wave circuit and MOEMS technologies have driven the optical system module into package optical system integration. At present, electronic, optoelectronic and photonic chips can only be integrated to produce a system with specific functions on the module level. More specifically, Polymer optical waveguide and electrical fuses are produced respectively on the PCB to achieve functional integration by welding the electronic and optoelectronic chips to the specified location on the PCB. Obviously, a variety of chips need to be integrated and assembled on a single substrate in the future—the low-light-wave chips. The theory behind low-light-wave chips dictates that we must first prepare the three-dimensional optical interconnection on the substrate by means of multi-layer optical waveguide interconnections, then weld the electronic chips and optoelectronic chips to the designated location via a flip chip bond process, and prepare an optical system in a package (OSiP). Such OSiP technology will accelerate optoelectronic and photonic processing technologies' ability to cover various application fields.

Related key technologies include: electro-optical simulation computing that can be integrated—optical DSP/CPU and other optical DSP system technologies of optical vector-matrix multiplication, core devices that can be integrated based on electro-optical simulation computing and processing technology such as vertical-cavity surface-emitting launcher (VCSEL) arrays, optoelectronic detector arrays, open space light modulators (OSLM) and other core optoelectronic devices, the optical interconnection on Polymer waveguide-based PCBs, light wave circuit substrates based on the multi-layer optical waveguide interconnections, micro-optical elements and MOEMS, and special CMOS chips for processing control.

**3. On-chip-integrated optical computing technology**

Since the 1980s, photon computers have become a research direction for new generation of computers. After 20 years of development with the integration of electronic and optoelectronic computing technologies, its enormous potential for upgrading computing speed has become more and more attractive, specially in the field of super computers.

Logic functions in optical computing have always been an all-optical control NOT, NOR, and NOT-XOR optical logic achieved by adopting the non-linear effect and cross-gain modulation in semiconductor optical amplifiers (SOA). In addition to chip integration difficulties in the above-mentioned approaches, its main problem is how to complete the above logic functions under low-light conditions. Studies have shown that in the near future it will be possible to achieve the functions of optical switching and optical transistors on the single-photon level—single-photon logic devices. The basic idea is to adopt controlled beams as the media by letting such beams pass through the surface plasmon polariton (SPP) generated at the metal/dielectric interface and have the interaction between the media and the neighboring two-level system (for example, single quantum dots) enhanced, by which signal beams can be controlled whether to penetrate the two-level system, thus achieving the logic functions of switching signal beams.

There are many schemes of storage in optical computing. The optical storage scheme for storing and transmitting mass data is mainly holographic technology. One classical option for optical cached storage is to adopt semiconductor ring lasers to achieve two kinds of light transmissions: from a single perspective, the clockwise transmission shows brightness as "1", while the anti-clockwise transmission shows darkness as "0". As long as the laser is appropriately injected to maintain a modest gain, memory function can be achieved. The quantum buffer memory for optical quantum computing commonly adopts the trapped ion scheme in the light trap. The quantum state of the signal light in the optical trap is stored by letting the atoms "entangle". The former light trap can be maintained for about 1 millisecond and will collapse due to the movement of atoms. Recently, scientists have used silicate crystals mixed with the rare earth element praseodymium to create a "super light trap." Because of its solid state and good magnetic stability, this trap boasts much longer time (to reach the second magnitude) to save optical quantum information than gas traps or unstable crystal traps. The trapped ion scheme in the optical trap will encounter great difficulties in meeting computer capacity enhancement and miniaturization in the future. Later it was discovered that no additional trap was necessary and optical storage could achieve millisecond magnitude at room temperature with a single spin entangled only by adopting the diamond nitrogen/vacancy defects.

Future optical computing that can be integrated needs to significantly reduce the chip area occupied by waveguide transmissions. By adopting surface plasmon polaritons (SPP) spread along the metal surface and excited by light, it is expected that light waves can be transmitted at the sub-wavelength scale (1/10–1/8λ) in the same way as in the past electrons were conducted through metal wires. Its transmission speed can be controlled as slower than, close or even higher than the transmission of the speed of light. The transmission can be 1,000 times faster while reducing its scale at the same time. On the other hand, surface plasmon polaritons (SPP) are also likely to carry quantum information and transmit entanglement as the transmission media between the quantum world and the outside world.

Clearly, the above trend has guided optical computing technologies that can be integrated to a new period of development: by means of CMOS chip technology, integrated optical computing chips are likely to be achieved at last. Related key technologies include: the core components of optical computing and processing technology that can be integrated such as all-optical logic elements and all-optical switches and transistors adopting the SPP effect; the technology for waveguide transmission and light speed control with adopting SPP effect; all-optical processors, associative storage and routers; classical optical cache and quantum optical buffers; massive optical information processing, storage and parallel transmission and other technologies.

## 5.4 Studying General Purpose Quantum Computers and Putting Them to Use

Quantum information is an emerging interdisciplinary field of quantum physics and information science. Based on characteristics of quantum mechanics, quantum IT boasts various new functions that cannot be achieved through existing IT such as: quantum cryptography that can provide the safe communications technology that cannot be tapped or deciphered, and quantum computing that can realize massive parallel computing equivalent to the exponential growth of the processing unit. Quantum information can provide new principles and methods for the development of information science, and it is expected to become one of the new generation information technologies in the Post-Moore era. The technology was born in the 1980s, having attracted the attention of international academics and national governments and experiencing a rapid development in theoretical and experimental research since 1994 when Shor proposed a quantum parallel algorithm.

### 5.4.1 The strategic requirements of quantum computers

Quantum computers are more powerful than classical computers in function, because the quantum superposition principle allows the simultaneous realization of a large number of computing, namely, the intrinsic parallel computing power. For example, the problem of integer factorization has found no polynomial algorithm in common computers, while with "quantum parallel computing", the problem can be solved by a polynomial algorithm (Shor algorithm) in probability. More great potential for quantum computing lies in its ability for quantum simulation. The accurate computing of multi-particle quantum systems can be realized only by means of quantum computers, and such simulation capabilities can help to identify the new phenomenon in meso-/nano-scale physics, and thus develop new types of devices and technologies. In order to develop the potential of quantum computers, the quantum mechanical system should be prepared and controlled under conditions far beyond the reach of current physics laboratories. So far, many quantum computer programs have been proposed, but several difference magnitudes are still waiting for solutions to problems related to the number of quantum bits or the number of quantum logic operations. Currently only a few problems can be solved by means of a small number of qubits. The foundation and technology are greatly challenging the genuine realization of quantum computers.

One serious obstacle to achieving quantum computing is the so-called decoherence problem. The problem states that the environment will inevitably destroy the quantum coherence of the system, enabling it to spontaneously evolve into a classical computer, completely losing the advantages of quantum computing. Researchers have found an effective means to overcome decoherence by using quantum coding theoretically. These programs use coded logic quantum bits to achieve robust or resilient quantum computing, and the cost is the introduction of redundancy in the number of physical quantum bits and basic quantum logic operations. It has been proved that, under certain assumptions, if each gate operation can achieve the threshold accuracy, quantum error correction can allow quantum computing to operate reliably.

As a macro-scale quantum system, quantum computers are bound to pair with the surrounding environment. In order to maintain quantum coherence, the smaller the pairing is, the smaller the impact of decoherence will be. Quantum computers, under ideal limits, should be closed systems completely isolated from the outside world to maintain their long-term quantum coherence. On the other hand, we need to be able to precisely control the evolution of quantum computers and read their results, and from this perspective, they should have an effective pairing with the outside world. Evidently, there are two conflicting requirements; hence both requirements should be satisfied when choosing what kind of physics system to create quantum computers. Frequently, coherent time is adopted to characterize the capacity of the physical system to maintain quantum coherence, the operation time to characterize the difficulty for external manipulation of the physical system, and taking into account both characteristic capacities is the maximum number of operations for computing in its coherent time. Clearly, all physical systems have their own advantages and disadvantages.

The past decade's research has shown that the realization of quantum computing is not difficult in theory, and that the bottleneck for current research is the physical realization of quantum computing. International academic circles have not yet have concluded what physical system may eventually successfully develop quantum computers, but it is generally agreed that solid-state-based physical systems and quantum-based optical systems are the most promising. Quantum computing of solid-state systems mainly includes superconduction and quantum dots, with the advantage of easy integration and the disadvantage of serious decoherence. Quantum computing of quantum optical systems includes ion trap, micro-cavity and atom chips, with the advantage of better quantum coherence and the main disadvantage of poor physical scalability.

The state of current experimental research:

(1) Unique features of quantum information have been demonstrated. For example, the stealth transmission of quantum states and quantum gate operations.

(2) Simple quantum algorithms have been demonstrated. For example, the Shor algorithm 15=3×5 has been successfully demonstrated in such systems as the nuclear magnetic resonance and photon entanglement.

(3) The accuracy of quantum coding theory has been verified through experimentation. For example, quantum error correction and decoherence-free sub-space have been verified in nuclear magnetic resonance, photon entanglement and atomic systems.

(4) There has not yet been a breakthrough in the physical realization of quantum computing, and especially in such areas as physical scalability and the quantum operation of low fault-tolerant thresholds, no substantive progress has been made.

The physical realization of quantum computers is not a distant dream. It has now reached the stage of basic experimental realization, and in the international arena a huge upsurge of research has occurred. It is not only necessary but also imperative to carry out research of quantum computing in China. The quantum computing can show its great power only when several quantum logic bits are reasonably under control; so, as a long-term goal, we should develop scalable quantum computing technology.

### 5.4.2 Scientific issues and key technologies of quantum computers

The strategic roadmap for the development of quantum computing boasts the overall aim to promote research in this field directing the development of scalability. "scalability" refers to: the capability for generating a sufficient number of physical quantum bits of memory to support logic encoding, and the capability for achieving qubit operations within the accuracy of fault-tolerant thresholds.

With respect to current scientific and technological developments, it is still difficult to prepare a physical system that can meet the above basic conditions. As a result, on one hand, we are making great efforts to look for a system that can meet the physical requirements while on the other hand, we are also exploring a new model of quantum computing to reduce the extent of demanding physical systems. This effort is in reality transferring the difficulty of preparing the hardware of quantum computers to difficulties in software preparation. For example, the distribution of the quantum computing model can effectively alleviate the harsh physical demands of physical scalability, but each distributed node must be connected into a quantum network by means of quantum entanglement, and it is necessary to achieve controlled quantum gate operations between different nodes. However, long-range quantum entanglement can be assured only with quantum repeaters, so the requirements for physical scalability in each node subsystem have been reduced, while the cost is to have a need for long-range entanglement and long-range entanglement operation. Again for example, the one-way quantum computing model can achieve quantum computing by measuring the single qubit, but in fact this model has transferred the difficulty to the preparation of a complex initial state (family state).

The output measurement of quantum computing is an important issue that differs from classical computers. In quantum theory, the measurement would interfere with the measured quantum sub-system and destroy the original quantum information of the system, causing a so-called "wave packet collapse", and the original quantum state, after the measurement, would collapse to certain Eigen state of the measuring instruments based on a certain probability. Quantum no-cloning theorem prohibits accurate copy of the original unknown quantum state, which has in turn created an insurmountable obstacle for extracting quantum information.

Quantum measurement refers to the process during which the quantum-bit states (i.e., information) are indicated in a classical system state over a period of time with a quantum-bit system under test and a coupled classical measurement system. One important index of quantum computing output measurements is: there will be a high probability to get the correct answer in the final state of good quantum algorithms adopted in quantum computing during measurements. For example, the Shor quantum parallel algorithm can correctly provide the function period.

Important research issues include: the research of quantum processors based on new materials and new structures, especially those systems with potential scalability; the experimental study of quantum coding; that is, the research of entangled logical qubits through experiments, which is the basic unit for further expansion of quantum computing systems; the research of "quantum quasi-processors" developed by a small number of quantum bits, which can integrate quantum features as well as classical controls and readouts into a single device; the research of the expansion and interconnection of quantum information storage units; the research of quantum simulation theory and programs to achieve the simulation that computers cannot achieve; the research of related support technologies, including the preparation of related materials and devices, electrical-optical systems, and single-photon detectors, etc.; the basic research of quantum computing theory, including a variety of new computing models, for example, topological quantum computing, etc.; research on the architecture of quantum computers such as what kind of structure is more suitable for scalable systems and how can we integrate quantum logic units and data storage as well as data transmission; the research of quantum algorithms; namely, studying what types of problems can be sped up for computing in quantum operations.

## 5.5 Beyond Zettaflops ($10^{21}$ flops) Supercomputing

### 5.5.1 The strategic requirements of super-servers

Super-servers, also known as high-performance computers, are the engines of the U-INS. A large number of super-servers will accept billions of service requests in the U-INS and run network service workloads. These computers are part of the critical infrastructure needed to achieve augmented value for the masses. By 2050, super-servers will be required to support various personalized application workloads and break through such technical barriers as low power, massive parallelism, reliability and low cost. Performance will need to increase by $10^8$–$10^9$ times in 40 years to achieve a computing speed of $10^{24}$ operations per second. During this process, major difficulties and technological breakthroughs will both occur in the transition phase from Exaflops ($10^{18}$ flops) to Zettaflops ($10^{21}$ flops).

Compared with developed countries, the greatest gaps in China's information industry are the server-side hardware, software and services. Super-server is an important part of the U-INS base platform. Supercomputing is also a strategic high ground for supporting cutting-edge technological development. Super-servers often lead in computer architecture and system software innovations.

### 5.5.2 Scientific issues and key technologies of super-servers

#### 1. Supporting the expansion and popularization of applications

From 2010 to 2050, traditional computer applications will be further expanded. Scientific and engineering computing, enterprise computing and personal computing and other new applications such as universal coverage, network services, data computing and intelligent computing will be integrated into the mainstream. The user group will not just include scientists, engineers and technicians, businesses or government, but rather all people including hundreds of millions of adults and tens of millions of students. The development of super-servers will have a particular focus on emerging application services and environmental constraints.

Computers' overall performance will increase by $10^8$–$10^9$ times in the next 40 years. Network services and low-cost terminals will deliver these capabilities effectively to mass users, which is bound to enhance existing applications and trigger a large number of new applications. Today's applications are only 10 percent of future applications. Computer science and technology will need to answer: How to characterize new application workloads? How to develop and maintain new application software and services? How to characterize new usage patterns scientifically? How to characterize user values scientifically and relate value to computer system parameters?

Defining values need to accurately characterize "network effects". How to measure the value received by individuals and by all users? What are the relationships between these values and system structure, cost and utilization? Computer science and technology needs to introduce new complexity metrics and explore the lower and upper bounds for various network service problems, induce fundamental workload phenomena and system phenomena, and sum up the necessary and sufficient conditions for network services to meet the important criteria.

**2. Breakthroughs in low energy consumption, massive parallelism, reliability and low-cost**

The design of traditional computers first targets performance. The performance of today's high-performance computers has reached more than one petaflop per second (Pflops), with energy consumption of more than 2 megawatts (MW), a degree of parallelism of more than 400000, the mean time between failures (MTBF) of less than 1 week, and a cost of more than 100 million dollars. Today's technical route is not sustainable, especially in the case of the reality where Moore's Law is slowing down. We need innovation in computer systems, keeping energy consumption, reliability and costs under control. At the same time performance will increase by $10^9$ times in the next 40 years and in 2050 operation speeds may reach $10^{24}$ operations per second (Yflops).

Achieve principle-based breakthroughs in low energy consumption. Reduce power consumption from system architecture, device technology, system software and power-supply control, with the energy consumption of a Zettaflops computer to be less than 10MW. Explore systematic integration of general-purpose computing components with special-purpose and reconfigurable computing components. Discover new principles and methods to deal with the complexity and imprecision to significantly improve the reliability of computer systems and achieve self-diagnosis and self-repair. Alter the direction of computer design to significantly reduce the cost of computers and strive to achieve the new Moore's Law with exponential reduction of software development costs. Achieve shift from high frequencies to multi-layer parallelism, substantially increase the degree of parallelism (up to $10^{15}$), and build up an efficient programming model. Significantly improve usability and intelligence of computing devices through breakthroughs in intelligent network services and intelligent human-machine interfaces.

**3. Revisiting the computer concept**

Due to the changes in application requirements and the emergence of networks, the concept of computers or computer systems has also experienced a shift. A network will connect all kinds of terminal equipments, servers, storage devices, peripheral devices, sensor networks and Internet of things to achieve the interconnection and integration of computing world, human society and physical world. The basic model of computing is developing from human-computer symbiosis to the human-machine-physical ternary universe. Computers are no longer just desktop tools, but are network computing systems including user clients as well as super-servers (high-performance computers) in data centers and computing centers. We need a new computer system architecture to depict and guide the multi-level designs of computer systems scientifically and systematically, including client computer systems, data center computers, Internet computers and human-machine social computers, etc.

With the integration of computers and networks, we need to reconcile the computer approach and the network approach. The essential features of the computer approach are: supporting high-level interfaces through virtualization and abstractions, and at the same time supporting direct execution of virtualized applications through system architecture technology. This is the key for computers to provide high efficiency in general purpose computing. The essential features of the network approach are packet-based multi-layer protocols stacks. At present, the architecture of computer hardware is still in very preliminary stages in terms of network support, viewing network as a common I/O device. Network applications and network services are low in operational efficiency and high in energy consumption. It is important to improve efficiency in designing new computer architectures and enable the computer to support dynamic, direct execution of network protocols and network applications.

Computers also include system software. System software innovations include operating systems, programming languages, compilers and debuggers that support massive parallelism; verification tools of large-scale programs; system software for datacenter computers, Internet computers and human-machine social computers; and self-repairing, self-optimization and self-management system software, etc.

## Historic trends of computer system performance

From the perspective of computing speed, both servers (high-performance computers) and client computers have experienced a historical trend of exponential growth with the growth rate increasing by $10^{13}$ times over the past 60 years and performance increasing by 1000 times roughly every 11 years on average. Development was slow from the 1960s to 1980s and it took 23 years to go from Mflops to Gflops, as breakthroughs in integrated circuits and parallel processing took time. Based on historical experiences, future application needs and scientific and technical challenges (circuit technology, low energy consumption, massive parallelism, high productivity and low cost), we estimate that there will be a period of overcoming challenges from 2015 to 2035, similar to the 1960s–1980s era, where major technological breakthroughs will be achieved, resulting in innovation and development in computer architecture, system software, application software, service models, as well as client-side high-performance computers and network computing systems. These breakthroughs will help to put power consumption and cost under control. It may take 20 years for the speed of high-performance computer to rise from Exaflops to Zettaflops. The performance of high-performance computers will have a further increase of $10^8$–$10^9$ times in the next 40 years, reaching $10^{18}$ flops in 2020, $10^{21}$ flops in 2040 and $10^{24}$ flops in 2050. The speed of high-end client computers will be up to $10^{12}$ flops in 2020, $10^{14}$ flops in 2040 and $10^{17}$ flops in 2050.

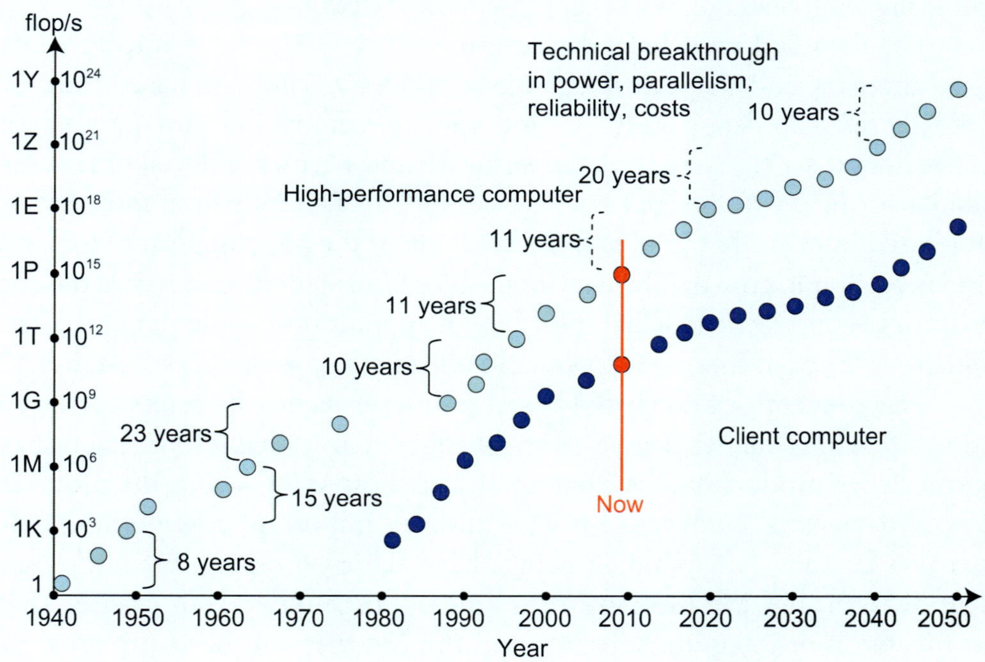

# 6 Developing the Data Knowledge Industry

The data and knowledge-based industry refers to the industry of producing data and knowledge, generating knowledge from data and creating economic and cultural value. Data and knowledge are also known as contents. Their consumption and production are influenced by culture and vice versa. The industry is rapidly developing and becoming an emerging economic market to promote industrial structure upgrades. The data and knowledge-based industry will be an important force in China's economic growth, and also the most promising component in market growth of the U-INS.

The data and knowledge-based industry can be divided into the data-based industry and the knowledge-based industry. The data-based industry generally refers to data collection, processing, dissemination, storage, analysis, management, services, and related software and hardware development and manufacturing. With the continual growth of China's information industry, the data-based industry has gradually become one of the most important forces in China's economic growth. The development of information resources is the core of China's economic and social information applications, while the data-based industry is the basis for development and utilization of information resources.

The concept of knowledge-based industries generally refers to certain industries with a higher degree of knowledge intensity based on the business of knowledge production, distribution and application as well as the provision of related services. Knowledge itself will be an important commodity in the next few decades. After human beings have achieved certain breakthroughs in understanding the human brain and intelligence, a leap greater than the invention of integrated circuits and the Internet will be required in the information industry. Laying the technological base for the development of knowledge-based industries will be a long-term scientific and technological task.

Data include structured, semi-structured and unstructured data, with their forms including spatial data, text data, voice data, image data, streaming data, etc. Data can be stored in a variety of media. In order to meet the high-capacity data storage needs of 1.2 billion people, we must carry out research on low-cost devices and systems with large-capacity storage. The Internet and information service industry will experience a bottleneck in computers' semantic understanding in the future. With the advent of the human-machine-thing ternary world, we must tap and utilize characteristics of Chinese civilization, research science and technology as well as the U-INS base platform, which can support semantic meaning, contents and culture in order to provide a scientific and technological basis for the development of knowledge-based industries with Chinese characteristics.

## 6.1 Target and Roadmap

Short-term targets (2010–2020): Develop low-cost front-end semiconductor storage and back-end mass storage technology. Develop large-scale parallel computing systems of low cost, low energy consumption, high throughput, high availability, easy programming and easy management, and support massive users in-time services of PB-level data. Study efficient data mining and knowledge extraction parallel algorithms, which can handle the data of time variation, multiple sources, multiple models and noisiness. Develop semantic Internet technology to realize data connectivity on the Internet. Develop practical Chinese language analysis tools, achieve speech understanding and scene understanding, provide specialized Chinese contents services and contents computing with users' participation, and explore the Chinese characteristics of data-knowledge services in cultural and creative applications. Develop the U-INS base platform supporting semantic understanding, contents and culture.

Mid-term targets (2021–2035): Develop emerging technologies including atomic memory and holographic storage to popularize storage services. Utilize the massive data from the human-machine-thing ternary world. Achieve knowledge and emotional understanding. Develop an intelligent Internet able to reason and learn. Explore and utilize Chinese cultural characteristics and Chinese peoples' production/consumption behavior. Develop a scientific and technological system of semantic, contents, and cultural data-knowledge services.

Long-term goals (2036–2050): Provide China's 1.2 billion users with ubiquitous storage capacity as well as personalized data and intelligent contents services. Achieve comprehensive understanding and multi-objective decision-making capabilities. Based on optimized mechanisms design, create a harmonious and civilized code of conduct and network culture with Chinese cultural characteristics, and establish security and trust between people and services (as shown in Figure 6-1).

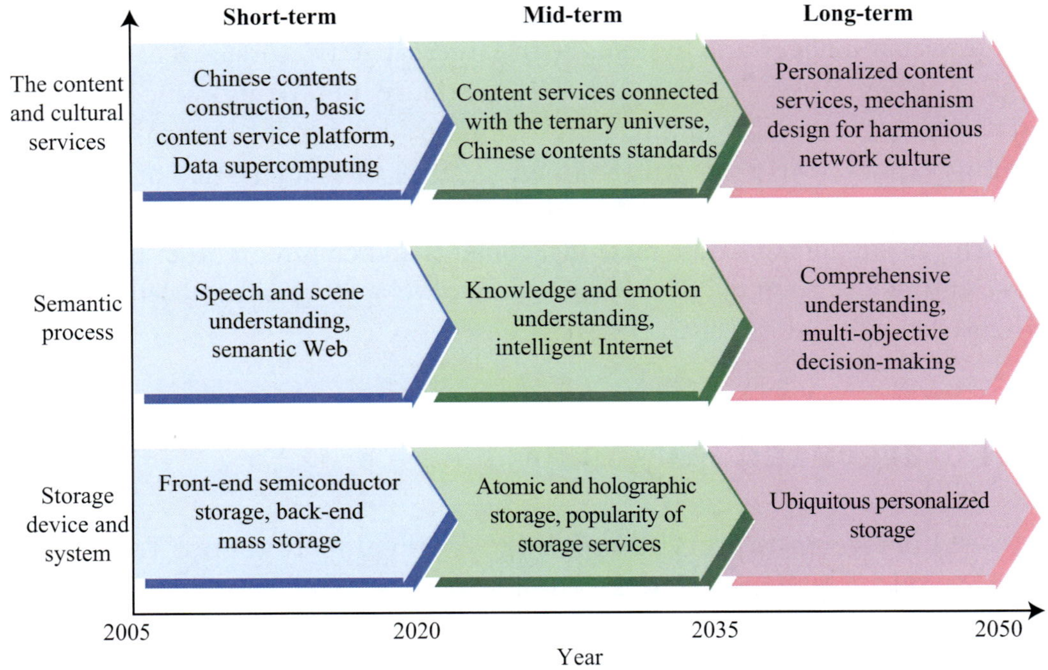


Figure 6-1 Roadmap for data and knowledge services

## 6.2 Ultra High Capacity and Low-cost Storage Devices and Systems

With the continuous development in breadth and depth of information, there is also a requirement for increased and more effective storage, while the retrieval of information will become a huge challenge. During networking visits, the demand for system storage capacity will be the primary force to drive storage technology development as long as basic access bandwidth, latency and reliability requirements have been met.

### 6.2.1 The development of new storage devices

**1. Advanced magnetic disk technology**

Although disk technology has experienced over 50 years of development, magnetic disks are still the primary storage devices used at present and will be for the near future. Current disk-capacity growth mainly depends on technological improvements in perpendicular magnetic recording (PMR). The latest technology has reached 600Gbits/square inch and it is expected that the annual growth rate will remain at about 40% for the next five years, and then around 2015 the density may reach 5Tb/square inch. The principle of magnetic recording is contingent upon closely aligned magnetic units, each of which is composed of a number of magnetic particles. The diameter of current magnetic particles has reached a level of 10nm, and further reduction will affect stability. Further development of the technology will require new technological breakthroughs. In addition to the introduction of new materials, there are two more kinds of research currently being conducted: one is Heat Assisted Magnetic Recording (HAMR), a method of using laser irradiation to heat, read and write the normal temperature storage; the other method involves the adoption of nanometer magnetic island technology to record information using single large magnetic particles (4nm or so). It is expected that a capacity of 50Tb/square inch can be achieved in about 2020, while magnetic disk technology and its applications have the potential to grow up to around 2030. Whether smaller magnetic recording units can be achieved in the future will depend on the development of atomic-scale technology.

### 2. Semiconductor memory

Semiconductor memory represented by flash memory has been developing rapidly in recent years, with a trend of replacing small magnetic disks in the low-end area. The biggest advantages of flash memory are its reliability, its limited mechanical structure, and its suitability for portable and microminiaturization applications. Flash memory technology will develop at its predicted speed according to Moore's Law, with processing technology currently at 50nm. Actually, flash memory density improvement and price reduction has been faster than that of magnetic disk memory in recent years. New technologies such as Phase-Change Memory (PCM) and Resistive Random Access Memory (RRAM) are developing rapidly. In addition, because the method of charge storage has been adopted, more precise control of a single memory unit can be achieved, and MLC has already been developed. Moreover, 3D flash memory technology is also being researched, and achieving multi-layered semiconductor memory is much easier than realizing multi-layered magnetic disk memory. In short, the memory density of semiconductor memory may gradually exceed that of magnetic disk memory. It is expected that in 5-10 years, flash memory may completely replace magnetic disks in the low-end area. But if flash memory is to replace the hard disks eventually, problems such as limitations in read and write times and stability will need to be solved. Current technology can realize a rate of one million read and write times, and is expected to reach more than 100 million times in the coming years.

It is predicted that by 2020, semiconductor memory will completely replace magnetic disks in front-end information systems. However, in the back-end mass memory area, semiconductor memory will be unable to replace traditional magnetic disk technology in the foreseeable future, since the mechanical positioning makes magnetic disk platters much larger than semiconductor chips, so the capacity is much larger as well.

### 3. Holographic memory

High-capacity storage technology that has the potential to replace magnetic disks will be based on optical storage technology. Light is an indispensable tool for micromachining because it can either target a more refined structure or replace mechanical devices to target a wide range of positions.

Currently there is more hope for holographic storage technology. Adopting the principle of optical interference, holographic memory technology records the interfered information in media similar to films. In present experiments, the recording density of holographic memory has exceeded that of magnetic recording. Another advantage of holographic memory is parallel random reading and writing, and the read-write bandwidth can be greatly increased. However, holographic memory has a long way to go from theory to practice. In fact, the concept of holographic memory has been proposed around for more than 40 years, so far it has not been put into practice, and proposed the current technology is barely on par with magnetic recording.

The application of holographic memory is restricted by the miniaturization of optical devices, such as CCD and LCOS technologies, and its foundation also depends on the development of semiconductor process technology. Holographic memory may first be used to replace existing Blue-ray DVD technology and along with the technological advances in semiconductor optical devices, it will gradually compete with magnetic recording and become a mainstream reality after 2020.

**4. Atomic-level memory**

Atomic-level memory is another kind of technology that is expected to replace magnetic recording in the future. It is conceivable that, with the continuous dimension reduction of carriers storing single information bits, eventually the smallest unit of memory may be a single atom. It may even be possible in the future to store multiple bits within a single atom. Recently IBM researchers were able to use five atoms to constitute a bit-storage with the diameter within 2nm. However, this memory needs to scan to tunnel microscope for atom movement and its read-write speed is extremely slow, so it is only of theoretical significance. It is expected that, when traditional magnetic recording technology reaches its limit by 2030, atomic-level memory technology may appear on stage as follow-up development technology and become mainstream technology by around 2050.

**5. Tape and backup systems**

As a backup storage device with a history of more than 50 years, tapes have begun to be withdrawn from use with small businesses as a real-time backup, and gradually replaced by CD-ROMs and hard disk-based backup systems. It is expected that small tape drives ceased to be used by about 2020. Mechanical devices of tape drives and tape libraries will be advantageous for use as large-capacity data backup. These devices have an aggregation storage capacity that can not be replaced by other technologies. The unit disk storage density development of magnetic tapes is correspondent with that of disks; therefore tape libraries will exist beyond 2030. By 2050, after new mainstream storage devices have been developed, replacement products for tape libraries will begin to appear, but it is expected that the last hierarchical mechanical structure will continue to exist.

**6. Associated memory**

Associated memory will be gradually developed in the future. Traditional memory is composed of many independent and passive access unit collections. This kind of storage will gradually mutate into associated and active storage. The device, combined with computing and memory including PIM and associative memorizers, will gradually expand the scope of applications, which is expected to be an important kind of product up to about 2020. The demand for memorizers integrated with retrieval and memory functionality will account for an important part of the total capacity up to 2035. A new class of associated memory devices based on optical holographic and complex macromolecules, etc., will appear.

## 6.2.2 Storage systems and services

**1. Unification of storage interfaces**

There are many kinds of storage systems that exist in the market today, including DAS, SAN and NAS and so on. The interface forms of future storage systems will tend to be unified, and can be generally divided into client-side storage and service-side storage. Client-side storage is mainly composed of portable devices and memory devices for local small-size caches, with limited capacity and fast access speed; while service-side storage is mainly concentrated on the data center. The service-side storage interface will be unified with future data networks, while large-capacity local magnetic disks used in personal computers will disappear. This process is expected to be realized by about 2020.

**2. The development of storage services**

With the further development of networking and software technology, the concept of storage devices will gradually disappear from daily life and be replaced by storage services. People will not need to face a multitude of storage media sizes, but instead be able to use the storage services of a single interface with reliable and unlimited capacity. The diversity of storage will be reflected in the service functions. A variety of new storage services in the follow-up phase will become part of the mainstream after storage service interfaces are unified. Traditional databases, search engines and other data processing software products will integrate with future intelligent storage services. It is expected that by around 2035, various storage services will become part of the mainstream, replacing simple data storage.

**3. Ubiquitous storage**

Even later in 2050, with storage technology gradually maturing, information in people's daily life will be automatically saved and have the ability to be searched on demand anywhere and anytime, without considering explicit storage operations. Storage technology will transit into a near-invisible background supporting technology.

## Resistive Random Access Memory (RRAM) and Memristor

In 2000, U.S. scientists reported that under the action of a nanosecond voltage pulse, the colossal magnetoresistive device can produce reversible changes between a low resistance state ("0") and a high-impedance state ("1"), i.e. electric-pulse-induced-resistance (EPIR effect). In addition, the resistance obtained can be preserved after the removal of the outer electric field. Based on this effect, scientists put forward a new concept of non-volatile memory - resistive random access memory (RRAM). RRAM is a brand-new concept that has become a new research hot spot in the field of physics and materials science since it was first proposed since 2000. The main advantages of RRAM include: simple preparation, with production based on existing semiconductor process technology and a significant reduction in development costs which can compete with NAND-type flash memory in terms of cost; high rewriting speed, generally less than 100 nanoseconds, which is much higher than flash memory; high storage density. RRAM is expected to become mainstream memory technology in the future.

Memristors are the fourth kind of passive circuit elements besides resistors, capacitors and inductors. The concept of "memory resistor" was proposed in 1971. On May 1, 2008, HP Labs confirmed the existence of memristors for the first time, and successfully created the first workable memristor prototype in the world. Memristors can match the capability of triodes invented 100 years ago, and any of their industrial applications have the potential to bring about revolutionary progress in the information industry. Compared with dynamic random access memory and flash memory widely used today, memristors have lower energy consumption and can store more data in circuits of the same size. The simplest application of memristors is non-volatile resistive random access memory (RRAM). One of the most interesting features of memristors is that all the "gray" states from 1 to 0 can be memorized, which is similar to the memory model of the human brain. Such hardware used in facial recognition technology can run a few thousand to several millions times faster than programs that run on existing digital computers.

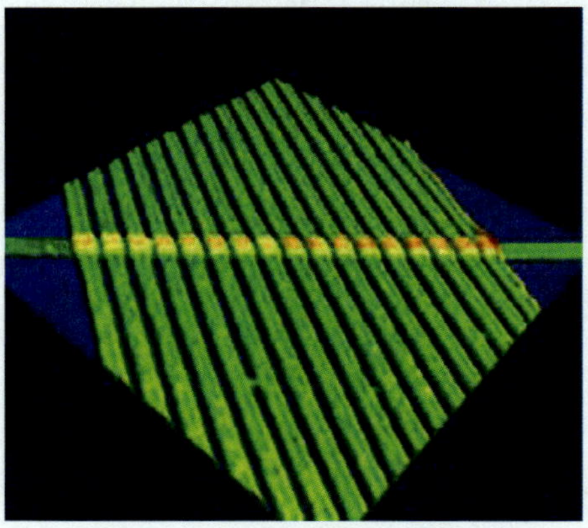

The memristor created by HP Labs
http://www.technologyreview.com/computing/20718/

## 6.3 Making Breakthroughs in Semantic Processing

### 6.3.1 Strategic requirements of semantic processing

Knowledge-based industry refers to those enterprises or organizations that emphasize producing knowledge, information products and services in particular, which must break through the difficulties of semantic processing. Data by itself make no sense; only data imbued with meaning can be used, and meaning of data is related to its semantic understanding. Semantic meaning is the interpretation of data symbols. It is also the concept of things in the real world as they correspond to data, as well as the relationship between these concepts and the interpretation and logic expression of the data in some particular field. For computer science, semantic meaning generally refers to the interpretation of computer representations (i.e. symbol) that users adopt to describe the real world; namely, it is the method employed by users to link computer representations to the real world.

Computer data are diverse in nature and include common text, graphics, images, sound, videos and animation, etc. The machine has to break through semantic processing if it wants to understand the content of these different media as human beings do.

### 6.3.2 Key technologies of semantic processing

#### 1. Natural language processing

Text semantic processing is essentially natural language processing. Natural language processing is not a general study of natural languages, but rather the development of the computer systems that can effectively achieve natural language communications, in particular, the software systems. Therefore, it is a part of computer science. The achievement of natural language communications between humans and computers means that computers should not only be enabled to understand the meaning of natural language texts, but also be able to express the intentions and ideas using natural language texts. The former is called natural language understanding and the latter is called natural language generation. Therefore, natural language processing generally includes the two concepts of natural language understanding and natural language generation.

Natural language processing, or more specifically, achieving natural language communications between humans and machines, or achieving natural language understanding and natural language generation, is very difficult. The root causes of such difficulties lie in the widespread variety of ambiguities or multi-meaning characteristics at all levels of natural language texts and dialogues. Generally speaking, most of them can be solved based on the corresponding context and scene provided. This is why we as humans usually do not feel natural language ambiguity and can use natural language to communicate effectively. In order to clear up the ambiguities, we need to adopt a lot of knowledge to reason logically. It is a difficult task for us to collect and sort knowledge more completely and find a suitable form to enter the data into a computer system and effectively use it to remove ambiguities.

At the current development stage, universal and high-quality natural language processing systems are still a long-term goal. However, for certain applications, practical systems with considerable natural language processing have emerged, and some have been commercialized, and even industrialized. There are some typical examples such as natural language interfaces of databases and expert systems, a variety of machine translation systems, full-text information retrieval systems and automatic abstracting systems.

**2. Image and video semantic generation**

With the development of image processing technology, multimedia technology and network technology, image information is becoming continually increasing in size, and how to automatically generate image semantics and understand image semantics is an extremely important key technology in image retrieval and pattern recognition.

Image semantics is divided into three levels, that is, the underlying feature level, the object level and the concept level. In essence, there is no semantic information involved in the image in the underlying feature level, and its basic idea is to extract the colors, textures, shapes and spatial relationships of the image. Content-based retrieval is accomplished at this level. The object level mainly factors in problems of the image object and the object space, extracts the perception features based on biological mechanisms and presents a framework model for learning many kinds of image semantics. The concept level is mainly used to solve meanings expressed by the image through the scene, knowledge and emotions. Scene semantics refer to the scene scenario in the image. Knowledge semantics refers to the reasoning of the knowledge in the image combining the scene semantics and the knowledge base, focusing on the act or action displayed by images. Emotional semantics refers to the understanding of the semantics contained in the images from the perspective of peoples' emotions, i.e. romantic images and horrific images. The semantic concept level usually relates to the abstract properties of the image, which is based on object recognition and underlying feature extraction, needs high-level reasoning for described objects and the meaning of scenes and objectives, completes the association between the underlying physical characteristics and high-level semantics, establishes the semantic concept-level index, and achieves the understanding and retrieval of semantic concept levels.

Existing video retrieval system technology primarily involves retrieval based on the underlying feature and non-semantic levels and is far away from the high-level semantic concepts understood by the human mind, which themselves would seriously affect the actual results of video retrieval. How to cross the gap between low-level features and high-level semantics and adopt high-level semantic concepts to create video retrieval is the key to the video understanding of computers. Efficient video semantic retrieval systems can be built based on a brief overview of semantic understanding, semantic analysis and semantic extraction of video contents.

**3. Speech recognition**

Speech recognition technology is a high-end technology that allows machines to convert voice signals into corresponding texts or commands through the process of identification and understanding. Speech recognition technology includes the extraction and selection of acoustic features, acoustic models and language models, etc. The extraction and selection of acoustic features is an important part of speech recognition technology. Acoustic feature extraction is a process for significant information compression, as well as a signal deconvolution process, in order to make model dividers better the partition.

Acoustic models involve computing from voices to syllable probabilities and language models involve computing from syllables to word probabilities. Language models can be divided into two categories including rule models and statistical models. Statistical language models reveal statistical laws within language units by means of probability statistics, and rule models refer to certain rules or grammatical structures in the language.

Due to the diversity and complexity of speech signals, current speech recognition systems can only achieve satisfactory performance under some limits, or can only be applied to certain specific situations. The performance of speech recognition systems largely depends on the following four factors: the vocabulary size and complexity for speech recognition, the quality of voice signals, the use of a single speaker or several speakers and the hardware platform. Speech act as the most natural communication media in the current communication systems, and with the development of computer and speech-processing technology, speech-speech translations between different languages will become the focus of speech research. Current research in the area of speech recognition focuses on: natural language database design; speech feature extraction; the research of acoustic model training based on language databases; research of acoustic models that have nothing to do with the speakers and adapt to the speakers; research of speech recognition algorithms; language translation and the research of speech synthesis and dialogue processing.

**4. Semantic Web**

The main task of semantic Web is to allow data to be automatically processed and understood by computers, so that computers can locate any information needed by people from the vast amounts of information on the World Wide Web, which in turn allows the existing information on the World Wide Web to develop into a vast global information and knowledge base. The main purpose of the research is to extend the current World Wide Web and enable the information in the network to have semantic meaning and be understood by computers, allowing for easy interaction and cooperation between humans and computers. Research will be focused on how to express information in semantic forms that can be understood and dealt with by computers. The development of semantic web technology is an important approach to achieving data interconnection on the Internet.

The underlying layer of semantic web is XML and RDF/RDFS, based on which we can build up ontology's and logical inference rules to complete the knowledge representation and reasoning on the basis of semantics, thus enabling us to be understood by computers. The RDF layer acts as the basis for data exchange, while the rule description language RIF and ontology description language OWL jointly act as the extension of RDF-S and describe the ontology and rule knowledge related to semantics. On the basis of unified logic, the goal is to realize certification through logical reasoning. Above the trusted web layer are the user interface and application layer expressing a variety of applications built on top of the semantic Web.

In order to enable computers to understand and deal with content on the World Wide Web, it is necessary to build up a large-scale ontology knowledge base in all walks of life, study unified logical reasoning and heterogeneous data mining methods, develop the technology of cross-media retrieval and hot topic positioning, automatically generate summary and comprehensive analysis reports, provide platforms and technology to achieve personalized emotional search engines for the World Wide Web.

## 6.4 Contents Computing and Culture Services

### 6.4.1 Strategic requirement of contents computing and cultural services

IT is a general-purpose technology. However, a developing trend of IT in the 21$^{st}$ century is related to diversity and local characteristics. Chinese culture is rich and has a long history. It is an important source to meet the growing economic and cultural needs of the mass users in China, especially in such areas as traditional culture, history, traditional Chinese medicine, architecture, ecology and online literature. Contents computing and cultural services offer a way to achieve the integration of Chinese culture and information technology. It is also one of the best potential channels for the U-INS to augment value for the masses.

Although the Chinese information processing industry has made great progress, the characteristic technologies used today mainly focus on such aspects as Chinese character input, character recognition, character classification and machine-based Chinese language translation. Furthermore, in comparison with English resources on the Internet, online Chinese resources are still very limited. There is much room for growth in Chinese contents processing and cultural services.

### 6.4.2 Scientific issues and key technologies of contents computing and cultural services

**1. Large-scale parallel data computing systems**

Develop large-scale parallel data computing systems (i.e., data-intensive supercomputing systems) with low cost, low energy consumption, high throughput, high availability, easy programming and easy management, supporting timely services for massive users. Provide China's 1.2 billion users with ubiquitous storage capacity and personalized intelligent data contents services. Current information network contents computing costs are too much and it consumes too much power. For example, to respond one user's search request, an Internet search engine needs to perform approximately O $(10^{11})$ computing operations and consume 1000–30000 joules of power. The search also usually returns hundreds of thousands results, most of which will not be exploited by the user. If the 1.2 billion Chinese Internet users search the Internet 10 times per person per day, then search engines will consume a minimum of 1.2 billion kWh of power in one year. There are still many algorithms and system problems needed to be solved, such as how to achieve low-cost, low energy consumption, efficient and accurate data services.

**2. Chinese content resource construction based on data and knowledge service platforms**

Content computing and cultural services should not rely solely on government promotion or businesses. Volunteer communities and the mass users are also value creators for content computing and cultural services. What the scientific community needs to do most is to provide companies, volunteer communities and users with value-creating platforms; or more specifically, develop a data and knowledge service base platform of neutrality and openness, with semantic/content/cultural support based on the U-INS in order to effectively support the acquisition, mining, management and value production of data and knowledge. This platform needs to effectively support users' participation and other business models, promote the development of various forms of Chinese language content resources, and integrate data and knowledge from the human-machine-thing ternary universe, in order to enable online Chinese language contents to increase to 10 percent of the world's total online contents.

**3. A scientific system for data and knowledge services with Chinese characteristics**

From the point of view of the integration of Chinese culture and information applications, explore and take advantage of Chinese cultural characteristics and Chinese peoples' production/consumption characteristics, and develop knowledge service systems that support semantics, contents and culture. Research basic concepts in data services, basic metrics and intrinsic laws, develop practical Chinese language analysis tools and provide specialized

Chinese content services and content computing. Make use of the penetration trend of information science and technology in the 21st century, to develop emerging disciplines such as computational history, network sociology, and data economics. Explore and refine the Chinese characteristics of data and knowledge services in such applications as cultural industry, personalized expertise information services and community information sharing.

**4. A intelligent Internet featuring reasoning and learning ability**

Develop semantic Internet technology to realize data connectivity on the Internet. Study efficient parallel algorithms in data mining and knowledge extraction that can handle time-variant, multi-source, multi-model and noisy data. Research reasoning and learning methods based on massive network data and users' involvement. Solve the noisy and high-dimensional data problems. Construct an intelligent Internet featuring a decentralized, developing and evolving knowledge base.

**5. Harmonious and civilized U-INS culture**

Create a harmonious and civilized human-machine-thing society with Chinese cultural characteristics. The current network culture in developed countries has utilized a combination of mechanisms such as laws and regulations, behavioral norms, netiquette, social assets and community pressures. A main feature is that culture is rooted in legal system measures based on contracts. Such contract-based culture could encounter problems such as rules explosion and high enforcement costs. Utilizing Chinese culture from the viewpoint of information science, with guidance of optimized mechanism design, we may be able to develop a network culture to avoid rules explosion and other modern problems, and achieve replicable information network code of conduct, solve reputation and trust difficulties, build up duplicable harmonious online communities, and enable the information network to become a civilized social space with both individual freedom and social harmony. Research on optimized mechanism designs could create a harmonious and civilized network culture, and build up trust between people and services. This will be an important way to realize personalized data and knowledge services.

# 7 Facilitating Industry Upgrade, Low Cost Informatization and Sustainable Development

## 7.1 Target and Roadmap

Short-term targets (2010–2020): Consolidate the IT applications basis. By 2020, the gap between the IT application level in the main areas of national economic and social life and that of medium-developed countries shall be reduced to 6–7 years from the current lag of more than 10 years; the general level of national information capability (the Networked Readiness Index) will rank 30$^{th}$ in the world from the current 50$^{th}$ to 60$^{th}$, and top five in Asia; China's eastern coastal areas with advanced economies will start to become elementary information societies.

The information industry will form an information industrial system that encourages domestic enterprises, is supported by self-created intellectual property rights, and has solid basic equipment capability. The information industry will form an initial sustainable development strategic pattern of low cost and high efficiency fitting the information industry of a large developing country. The added value of information and communication technology (ICT) industry will account for more than 9 percent of GDP. A national information management system and cross-department information resources sharing system will be established. Informatization will gradually penetrate every aspect of the national economy, and traditional industries will undergo large-scale transformations based on informatization. The informatization gap between urban and rural areas as well as between the eastern and western regions will be narrowed. Informatization will become an important means to raise people's knowledge cultivation and quality of life, and an important means and strong driving force to strengthen the building of a spiritual civilization, improve the quality and level of education, and build and promote the construction of a harmonious society. A social information service system targeting all of society will be in its initial stage, with significant improvement in public health, job security and social management. The concept of e-government will be fully implemented and the government's macro management functions will have started a transformation, relying more on information-based environments.

By 2020, China's desktop office and learning systems, servers and communication network equipments based on user controllable technology shall attempt to account for more than 50 percent of the domestic market and create the foundation for significantly reducing the cost of informatization.

Mid-term targets (2021–2035): Promote informatization comprehensively. By 2035, the application level of information technology shall be significantly improved, and the gap between China's economic and social informatization levels and that in moderately developed countries shall be shortened to two to three years; China's Networked Readiness Index shall rank among the top 20 in the world and top three in Asian; the level of economic and social informatization in China's eastern coastal areas, which is economically developed will approach and catch up with that in moderately developed countries. China begins to fully transform into a modern information society.

Major domestic telecommunication services and manufacturing enterprises will have become multinational companies among the international forefront. A long-term service mechanism with the core of a universal service fund will be established, laying a solid foundation for the balanced development of information and communication services in rural and western areas of China and gradually expanding the contents of universal services. The added value of the ICT industry shall account for more than 12 percent of GDP. Comprehensively establish information management systems and a sharing system for all department information resources. A management system will be formed facilitating civil-military integration.

By 2035, China's information products and services based on user controllable technology shall account for more than 60 percent of the domestic market and allow informatization costs to be greatly reduced.

Long-term targets (2036–2050): China's economic and social informatization will be highly developed and will catch up with the level of moderately developed countries. China Networked Readiness Index shall rank among the World's top 10. China will become an intermediate information society. The economically developed areas on China's eastern coast will become an advanced information society.

By 2050, 800 million computers will be in use and the number of Internet users will reach more than 1.2 billion (including mobile Internet users) with the Internet penetration rate reaching over 80%. Information processing capability will be almost unlimited and quantum cryptography will make information systems secure. Governmental regulations and personal freedoms will be balanced properly and personal privacy will be highly protected. China will form informatization patterns suited to modern countries and join internationally competition. The information skills of the masses shall be markedly improved. Information technology shall extend to all walks of life in the national economy and all aspects of people's life, becoming an assured strategic platform to improve quality of life and build a harmonious society (as shown in Figure 7-1).

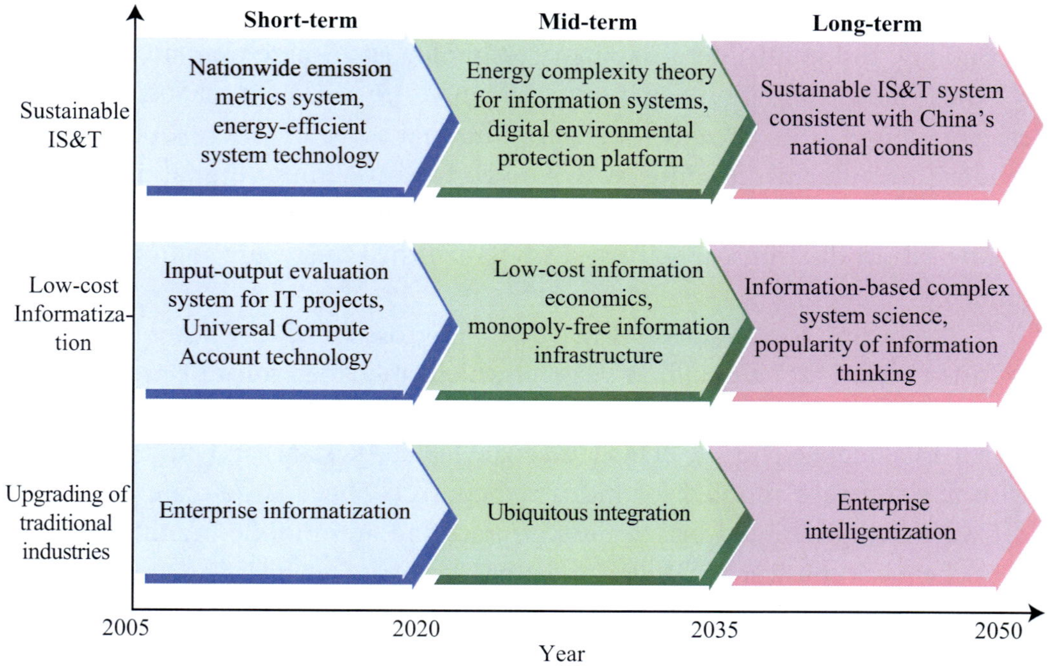


Figure 7-1 Roadmap for industrial upgrading and a low-cost informatization

## 7.2 Developing Industrial Software and Upgrading Traditional Industries

### 7.2.1 Strategic requirements of industrial software

The process of industrialization has offered opportunities for the naissance of industrial software. The first wave of industrialization focused on mechanization, the second on electrification and the third on automation. With the invention and wide application of integrated circuits, industrialization has entered the fourth wave primarily characterized by digitalization, which is typically represented by CNC (Computer Numerical Control) machines and systems. Widespread applications of networks have promoted all phases of the product life cycle, such as research and development, design, production, testing, supply and marketing, to a new level ready for a brand-new integration stage of informatization and industrialization. Clearly, industrialization is entering its fifth wave, where industrial software will play a special role in upgrading the operations and managements of various companies and enterprises.

Industrial software is the fusion of informatization and industrialization.

It is the core technology allowing the production equipments featuring mechanization, electrification and automation to be further featured with digital, networked and intelligent applications. It integrates product design, manufacturing, testing and management, and heavily uses simulation technology and multi-interdisciplinary technologies. If there were no industrial software, industrialization could only remain at the level of mechanization and electrification. If no breakthrough in industrial software, there would be no modern manufacturing industry as known today. Without digital, intelligent and network-based technologies, it would be impossible to eliminate the backward production capacity and revitalize the equipment manufacturing industry.

According to the statistics in 2008, the gross value of Chinese software industry was about 757.3 billion Yuan, of which embedded software accounts for about 111.8 billion Yuan. In China, embedded software was mostly adopted by digital TVs, mobile phones, PDAs (Personal Digital Assistant) and other products, with few software adopted for industrial control. Those large-scale industrial software being adopted by Chinese aircraft and automobile manufacturing, petrochemical, construction design and other industries rely mostly on import. The price of a set of software varies from several million US dollars to as high as several ten million US dollars. In recent years, western countries have begun to restrict the exports of industrial software to China, such as oil exploration software. Compared to operating systems and common software, Chinese industrial software technology falls even further behind the western countries.

As the soul of the embedded computing systems, embedded software is an important part of industrial software. Thus the development of the embedded software industry becomes a breach point for the transformation of Chinese information industry from "Made in China" to "Created in China." It is also an important way for Chinese information industry to change from an extensive growth mode to an intensive one, and achieve sustainable development. R&D capabilities of embedded systems and their industrialization level have become an important symbol to measure a country's economic development, science and technology progress and national defensive power.

Industrial software is very complex, and its design is as difficult as the design of large-scale integrated circuits. The development of industrial software requires the interdisciplinary cooperation and the collective efforts of scientists, engineers and technical experts. The development of industrial software first requires mathematical modeling and algorithm design, rather than software programming. Therefore, we should devote major efforts to the development of modeling techniques, algorithm design and optimization techniques, as a good mathematical model and an ingenious optimization algorithm can often promote the development of a whole industry. In addition, the development of industrial software is not just the development of software products itself, it is more important to realize the services of industrial software.

According to the statistics from 1870 to 1980, the efficiency of the production process increased 20 folds, but the efficiency of enterprise management only increased 1.8–2.2 folds, and the efficiency of product design only increased 1.2 folds. It is thus evident that improving the efficiency of both product design and enterprise management will be the development focus of industrial software in the future. According to the historical experience in the U.S., physics-based industrial software triggered a revolution of high-tech manufacturing products, which was a major reason why the high-tech manufacturing products of USA became so competitive. This revolution will last several decades to complete. To create and promote an industrial software product, an interdisciplinary team with more than 20 experts will take five to ten years to focus on the work continuously before the software product can reach its mature level of massive production application. By 2050, China will need tens of thousands of such advanced industrial software. The development of the scientific industrial software is an important way to strengthen Chinese industries, to significantly improve Chinese industrial structure, to expand high-tech industrial markets, and create more high-tech jobs.

The mid-short-term goals of Chinese industrial informatization include transforming manufacturing industry with informatization, promoting the informatization of product design and R&D, digital production equipment, intelligent production processes, networked management, and enterprise transformation, making full use of information technology to drive the transformation of those industries characterized by high energy consumption, high material consumption and high pollution, promoting Supply Chain Management (SCM), Customer Relationship Management (CRM), and vigorously support informatization of small and medium enterprises (SME), making great efforts in the e-commerce development to reduce the costs of logistics and transaction costs, improving the informatization level of mechanical and electrical equipment to achieve more accurate and efficient production, promoting distributed control, field bus control, agile manufacturing and other technologies, and enhancing the on-line monitoring, early warning and control of the production processes.

From the perspective of medium to long-term development, industrial software will promote the transition of traditional industries from informatization to integrated and intelligent applications. Based on the achievements of informatization, the integrated application will promote the enterprise transition starting from internal full life cycles integration (including the design, production, sales and services) and all-round levels integration (including device control, manufacturing execution and enterprise management), to the whole industrial chain integration. The networked virtual manufacturing based on Internet platforms will gradually become the dominant mode. Intelligent application will promote enterprise from information integration and functional integration to knowledge integration and intelligence integration, and ultimately realize the integration of human-machine intelligence in the enterprise; namely, the organic combination and fully utilization of human intelligence and software intelligence.

### 7.2.2 Theoretical and technical system of industrial software

#### 1. The theoretical system of industrial software mainly includes the following

**(1) Developing the innovative theory of product design.** This is aiming at the improvement of product design efficiency and quality in order to meet the requirements on customization, humanization and green manufacturing. The main content includes: digitalization modeling, simulation evaluation theories and design specifications of product's full life-cycle; design, planning and management techniques of rapid product innovation and development; theories and methods of end-users' design of their own products; theories and methods for web-based collaborative design of complex products.

**(2) Integrated modeling and optimization theory for various kinds of flows.** Enterprises involve various dynamic flows, such as material flow, energy flow, information flow, capital flow, value streams and so on. For example, process simulation is mainly concerned with the emulation of material flow and energy flow, which can be used for new process design, new equipment design, and optimization control of production equipment and processes, etc. Similarly, if we can grasp the operation laws of those flows themselves and the interaction laws among various flows, we can clearly understand the development and the change laws of the overall enterprise, and continuously enhance enterprise management efficiency and quality. The main research will focus on the integrated modeling theories, integrated optimization theories and a variety of core algorithms for various types of flows.

**(3) Innovation theory and parallel management theory of enterprise management.** The goal is to continuously improve the enterprise management efficiency, promote the great-leap-forward development of the enterprise operation and management, change the enterprise management mode to adapt to the development needs, speed up the process from integrated applications of informatization and industrialization to intelligent development. The primary research will include:

- Technologies and methods for the software realization of comprehensive enterprise innovative theory: Enterprise management based on globalization and integration will move from sole or partial innovation management to the comprehensive innovation management of the whole industrial chain. The main research will focus on technical innovation, product innovation, production innovation, service innovation, management innovation and strategy innovation, etc.
- Parallel control and parallel management theory: we can establish artificial systems for the "equivalent" description of actual complex

dynamic enterprises by using relevant theories and technologies, and understand various change laws and interaction laws between elements of enterprise complexity and social complexity by means of computational experiments. On this basis, we can realize the parallel control of production device level, parallel management of production and enterprise-level through artificial systems and actual systems (enterprise information systems), that artificial systems can enhance actual enterprise management level, prevent the accidents under normal conditions, as well as improve the emergence processing speed and quality and reduce the loss under abnormal conditions. The main research will focus on the core theories, key technology and software development of artificial systems, computational experiments, parallel control and parallel management.

**(4) Enterprise intelligence theory.**

- The enterprise collaborative strategy theories and key technologies in the network environment. Both social and engineering factors at different levels, different aspects and different stages of enterprises internally and externally require organic integration. The Interaction laws between the enterprise elements are increasingly complex. The constraints and objectives needed by enterprises have become stricter and stricter. The main research will focus on the enterprise collaborative strategy theories with multi-agent, multi-constraint and multi-purpose in the dynamic network environment, and the integrated modeling, analysis and optimization theories.
- Intelligent control and management theories and technology. In order to improve the intelligence of industrial software, namely its "IQ" and "EQ", the main research will focus on data mining techniques, knowledge representation techniques, and intelligent analysis techniques to provide theoretical support for intelligent control, intelligent management, enterprise intelligence and business intelligence software, etc.
- Theory and technology of human-machine intelligence integration. The ultimate goal is to achieve the organic integration of human intelligence and software intelligence. The main research will focus on the theory and technology of expression, processing, modeling, analysis and utilization in the areas such as human behaviors, languages, knowledge and wisdom (learning, analysis and decision-making); human-machine integration theory and technology based on the organic integration of artificial intelligence in industrial software and enterprise staffs' collective wisdom.

**2. The technical system of industrial software**

**(1) High-performance computing platforms and technologies of industrial software.** Industrial software is becoming increasingly large, complex and intelligent, which means that data, knowledge and intelligence will need a rapid increase in storage capacity, computing processing capacity as well as networked hardware platforms. The main research will focus on industrial applications of high-performance computing platforms and technologies; technologies of transforming enterprise data center and information center into knowledge center and even wisdom center; high-performance universal parallel computing technology, dedicated parallel computing technology, cloud computing technology, and grid computing technology of industrial software.

**(2) The interconnection and integration technologies and standards for industrial software.** The main research will focus on the network technology to provide more and more powerful wired/wireless sensing, to connect information sources from enterprise internally and externally as an organic whole in real time, and on the interconnection and integration technologies and standards of various kinds of information systems on network, hardware and software layers.

**(3) The architecture and platform technologies and standards for industrial software.** Industrial software has become increasingly large and complex. Its new progression and success depend largely upon its safety, usability and reliability. Therefore, the main research will focus on the architecture and platform technologies as well as the standards for industrial software. They mainly include: SOA (service-oriented architecture) for large-scale industries; SAAS (software as a service) framework to meet SME's requirements by using "long tail theory"; Web 2.0 and Web 3.0 to Web n.0 technologies to meet service trend of industrial software development; industrial software platform technologies and standards to realize information integration, functional integration, process integration, knowledge integration, wisdom integration and human-computer intelligence integration.

**(4) Languages and tools of industrial software development and configuration.** Both the formation of an advanced industrial software industry and the development of efficient and high quality industrial software require studying advanced industrial software languages, development tools, configuration tools, testing maintenance tools, infrastructure components, middleware and visualization tools, etc.

## 7.3 Realizing Low-cost and Effective Informatization for the Masses

### 7.3.1 Strategic requirements of low-cost informatization

As we enter the era of information economy, information has become an important resource for economic development. Economic activities consume resources. Information consumption incurs costs. With the development of the information economy, information costs are also growing. In 1999, the informatization investment of U.S. multinational corporations accounted for 3.5 percent of sales, and increased to 5.8 percent in 2005. During the Eleventh Five-Year Plan period, China's SME informatization investment is expected to exceed two trillion Yuan. We must strive to ensure the efficient allocation and returns of such huge investment. In order to study the economic benefits of informatization, information economics was born in the 1940s and became a formative study in the 1970s. Information economics is a science that intersects economics and information science in order to optimize the value of informatization through scientific organization, rational optimization, precise controls and the timely adjustment of human information activities.

Over the past 20 years, IT has been in a period of rapid growth. Major research and development of various companies all focused primarily on improving function and performance. Moore's Law reflects the expectation in terms of technology development of large international semiconductor enterprises. China's informatization process also experienced a range of issues, such as high investment, long cycles, rapid updates and unsatisfactory effectiveness. In the 21st century, informatization has begun to enter a phase of mass adoption. Users are not just a small number of IT enthusiasts or the wealthy class, but also hundreds of millions of ordinary people; hence low cost, ease of use and security should be given more consideration than the performance. China should emphasize research and development of low-cost information terminals, networks, servers, as well as low-cost information services, to enable the country's 1.2 billion people to become IT users. China must find a way of low-cost informatization in which the national condition is a specific concern.

Low-cost informatization is contingent upon China's domestic conditions. First, due to its vast territory, the construction and renovation costs of the information infrastructure in China are much higher than that in most Western countries. Second, China has such a large population needing a large number of terminal devices for the masses. In addition, China has not yet completed the process of industrialization; hence industrialization and informatization should be carried out at the same time, and we should promote industrialization with informatization to achieve integration of the two. More importantly, because China is still a developing country, there exists serious digital divide as a result of regional differences and differences between urban and rural areas. At present, the average household computer ownership rate in cities reached 60 percent, and even 90 percent in some developed areas (such as Beijing and Shanghai), meanwhile the rate is only 3.6 percent in rural areas. In the process of informatization for the masses, we must strive to reduce the digital divide with particular attention paid to less developed regions and vulnerable groups. Such strategic requirements present a great challenge to scientific and technological professionals in China. We must establish an information economics adapted to China's national characteristics and a cost-effective information infrastructure base unlike that of most developed countries. We must do our utmost to reduce the total cost to achieve informatization for the masses and truly allow the general public to benefit from informatization.

Achieving cost-effective informatization and establishing user controllable information technology and information service platforms are inseparable activities. Currently, competition among information products and services on the market is essentially the competition among platforms. Low-cost informatization can not be achieved based on monopolized platforms of foreign enterprises. For example, core technologies of the current desktop office applications and mobile phone communication systems are controlled by a few monopolies; therefore the cost of building the information infrastructure can not be significantly reduced. Low-cost informatization can be achieved only if a user controllable information platform based on community-created intellectual property rights technology is established.

Low-cost informatization shall not be achieved at the cost of reducing effectiveness and value, and value should be proportional to popularization. Informatization for the masses does not only mean an increase of the user number enjoying the minimum level of value, but also means that users enjoying higher levels of value are also increasing, thus promoting healthy and sustainable development of the IT market. Since China's reform and opening up, the disposable income of China's urban and rural residents has been increasing year by year and in the decade of 1998–2008, the average annual income of residents increased by 9.4 percent. This trend has laid the economic foundation for an increase in national propensity for information spending. We should fully recognize the rise and strong demand of China's "digital natives." According to statistics in 2008, among the 300 million Internet users in China, 67 percent of them are younger than 30 years old and 32 percent are students, whose information requirements in the next 30-40 years are far above the current information poverty line level.

### 7.3.2 Scientific issues and key technologies of low-cost informatization

From the perspective of end users, IT costs include transaction costs, cognition costs, purchase costs, use costs and maintenance costs. In China the key to achieve cost-effective informatization is to achieve core technology breakthroughs, build a low-cost informatization base platform, and carry out informatization applications appropriate to the conditions of China. For this purpose, we must address the following scientific and technology issues.

**1. Low-cost informatization calls for value augmenting mass adoption**

Low-cost informatization does not mean low value. It is much more than providing 1.2 billion people with an Internet-accessible cheap computer. It implies value-augmenting IT for the masses. The table at the end of Chapter 2 shows three forecasts considering the value-augmenting growth route, and the actual data comparison with 2000 and 2008. Based on the conclusions of the

*China Modernization Report 2006*, by the mid-21st century, China's productivity will reach the level of developed countries in the early 21st century. The pessimistic estimation expects that by 2040, the total IT users in China will reach 1.2 billion without changing the proportion of various groups of users (namely, the digital divide will not be narrowed). The medium estimation assumes that there will be a marked improvement; that is, all users will be above the information poverty line (the digital divide will be narrowed, and all users will be free of information poverty). The optimistic estimation assumes that there will be two more significant changes: The information consumption per capita will reach the level of developed countries in the early 21st century; The GDP per capita will coincide with the estimate of the 2008 Carnegie Foundation Report.

The above three predictions all demonstrate two important conclusions: only value-augmenting growth can be regarded as an effective low-cost informatization route; and today's information market is only a small fraction of the market in 2040. During the period from 1998 to 2008, the annual growth rate of China's IT market was 14.5 percent. Even at a conservative estimate (that the annual growth rate of the IT market will only be 4.75 percent), China's IT market will still quadruple by 2040.

**2. Establish information economics corresponding with China's national conditions**

To solve the problems such issues as how to promote industrialization through informatization, how to achieve the integration of informatization and industrialization, and how to bridge the digital divide, all require a set of information economics necessarily both in line with China's national conditions and the IT development trends. Information economics with Chinese characteristics is an important component in the concept of scientific development. In particular, the in-depth research of network effects based on IT will be carried out to enable network effects to surpass Moore's Law and become the guiding principle of value creation, cost reduction and universal access to IT. Whether information construction is successful shall ultimately depend on the level of input-output ratios; hence we should study an evaluation system for informatization consistent with scientific development concepts under the guidance of information economics, relate information costs to service value, and evaluate the informatization process on the basis of the general public's satisfaction.

**3. Establish a user controllable information technology platform**

We must establish a user controllable base technology platform in network terminals, servers, network communications and information services, lifting control of key technologies from monopolies. In particular, we should conduct research and development on a group of low-cost key technologies appropriate to China's national conditions. We must also strengthen the exchange and cooperation with international scientific circles, actively participate in volunteer communities in developing open source codes and open standards, and set up an innovative ecosystem of IT characterized by openness, opposing monopoly and win-win cooperation.

**4. Develop information-sharing technology**

The information industry is characterized by high costs for production and extremely low costs for reproduction, therefore an important way to reduce the informatization cost is to achieve maximum information sharing and reuse, build up effective economies of scale and better utilize information systems. Cloud computing and grid services emerging in recent years are leading development directions towards information sharing and cooperation. Breakthroughs in information-sharing technology can best reflect the intention of informatization. The rapid development of data and knowledge services in ubiquitous information networks is an important means to promote and achieve information sharing. In particular, we should pay more attention to the new and emerging information service market rather than being restricted by compatibility with existing information products and services.

**5. Eliminate the negative impact of information product upgrades**

Improve the reliability and easy maintenance of information products and eliminate the negative impact of information product upgrades. The total cost of ownership of information products is much higher than that of acquisition cost; hence we must do research on IT featuring high reliability and easy maintenance in order to significantly reduce the total cost of ownership. We should study new information system architectures featuring friendly upgrades to make end products basically free of upgrades, or make upgrades to have no negative impact on users. We should also change focus of the blind pursuit of high performance and emphasize the performance-cost ratio, the performance-power ratio and the value-cost ratio.

**6. Develop universal compute account technology**

Provide a universal computing account to all of China's 1.2 billion users, enabling users to conveniently and efficiently utilize their information environment including online resources at any time and any place, via any device. This account will be personal and decoupled from the information terminal devices and network service providers. The personal information environment shall be no longer bound to information terminal devices and network services and as such will be able to achieve human-oriented information services.

**7. Create a flat information infrastructure and substantially increase operating efficiency**

According to international survey data, the current operating efficiency of servers and data centers is only 10 percent, but users can no longer add more workloads. The main reason is that the information infrastructure is complex in nature, featuring many layers of network services, high system jitters, low scheduling efficiency, etc. We need an information infrastructure simple in structure to enable information applications and services to execute directly and efficiently, and to raise operational efficiency to over 60 percent. A key is to solve the conflict between computer system architecture and network layers, and to achieve breakthroughs in the computer architecture technology that directly supports dynamic network services.

**8. Develop information-oriented complex system science and simplify value realization**

Low-cost informatization systems generally involve network computing systems. These are complex systems. But a complex system does not always mean complexity in use. We need to study the science and technology that can positively embrace the complex system phenomena as an advantage, to simplify operations for end users. Specifically, we should study technologies that utilize data-, information- and knowledge-oriented emergent phenomena, as well as links, clustering, intelligent interfaces and intelligent processing, to hide nonfunctional complexity of IT systems and significantly reduce the users' cognition and operation costs. For example, there exists a phenomenon of "zero ten-thousandth (0/10000)" in current web searches, which means that ten thousand web pages are returned after a search without any desired result. However, initiatives such as Linked Data focu on the goal of "ten tenths (10/10)."

**9. Promote computational thinking to the masses**

Refine the discipline structure of computer and information science and technology via computational thinking, and change the social impression that computing is only a tool. Through 40 years' efforts, computational thinking will penetrate various fields of society. By 2050, the vast majority of users in China would have computational thinking ability.

### Two Ways for China to Realize Low-cost Informatization

The effectiveness of informatization can be measured by expenditure per capita on IT (including computer hardware, computer networks, software and services). This metric can be grouped into five grades qualitatively and quantitatively. The following diagram quantitatively characterizes two for low-cost informatization. The path of poverty line growth may popularize IT among 1.2 billion people, but the per-capita value will be low and the market will shrink, making it an undesirable growth path. The path of value-augmenting growth is a more desirable way for the low-cost informatization.

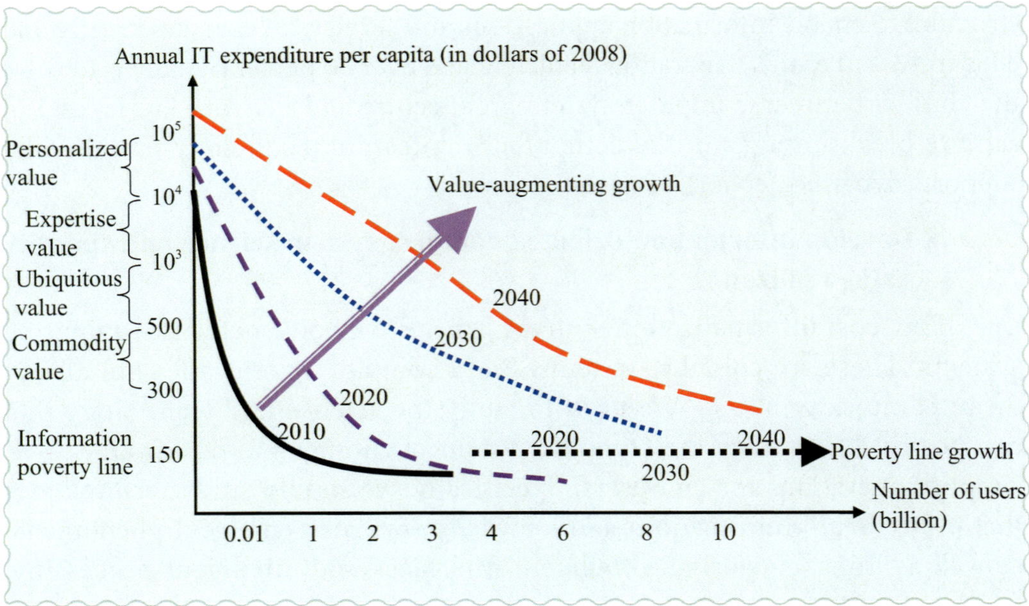


## 7.4 Realizing Sustainable Development of the Information Industry

### 7.4.1 Strategic requirements of the information industry for sustainable development

By 2050, China's Internet users will exceed 1.2 billion and the information industry market may be ten times larger than the present. Informatization will be the backbone of the national economic and social development. We must consider how to achieve the sustainable development of IT. In the next few decades, we will face a great challenge of how to significantly reduce energy consumption, resource consumption and emissions caused by information products and services.

The development of IT itself must take the path of sustainable development. IT must adopt a technical route of low energy consumption and low emissions. At the same time, IT must also provide technology support for sustainable development of the information industry. The current information industry is not a green industry, but an energy-intensive one and a major polluting source. Only if IT itself becomes a sustainable industry can it support the sustainable development of the economy and society. However, the present information industry has not yet reached the stage of sustainable development.

### 7.4.2 Scientific issues and key technologies of the sustainable development of the information industry

**1. Formulate realistic and sustainable development goals**

It is a necessary prerequisite for achieving sustainable development of informatization to formulate scientific and feasible goals, which in turn requires that we develop scientific and meticulous methodology and databases, and establish precise causal relationships aimed at reducing energy consumption, resource consumption, and environmental impact (such as carbon emissions), and their relationships with information value. Based on empirical evidence, China will be able to achieve the goal of a consumption emission growth rate significantly less than the GDP growth rate within 25 years and a zero growth in consumption emissions within 40 years.

The process for achieving sustainable development of China's information industry can be divided into three stages. The first stage (2010–2020) will be to significantly decrease the power consumption growth rate in the ICT industry, while maintaining the continuous growth of the ICT market. In the second stage (2020–2035), Integration of industrialization and informatization shall be achieved. The growth rates of power consumption and emissions in information service and communications industries will be significantly lower than that of China's GDP. In the third stage (2035–2050), the overall level of China's informatization will surpass that of moderately developed countries, and the goal is achieved to have zero growth in energy consumption and pollution emissions in the ICT industry.

**2. Explore the transformative S&T of low energy consumption, low resource consumption and low emissions**

In today's computers, each ALU operation consumes energy of Pico-joule magnitude at the circuit level and of micro-joule magnitude at the system layer, both much higher than the theoretical lowest limits offered by Physics. There is much room for reducing energy consumption in information systems. Low-power technology involves such aspects as materials, devices, systems architecture, and system software and management models. Principle-based innovations and breakthroughs in the core technology of low power consumption will become major challenges for chip and system design in the coming decades.

**3. Establish a computer-supported national economic sustainability indicator system**

This area focuses on precise statistics as well as computing and analysis of energy-saving, emission-reduction and balanced targets. Its goal is to establish an accurate information system in terms of national resources and energy, focusing on precise surveys and the tracking of China's major energy resources, and answering such questions as how much resource is used, where they are used and where bottlenecks occur. Addressing the strategic requirements of China coping with global climate changes, we need to firmly build up carbon cycle computational science and carbon cycle computational economics.

**4. Develop the sustainable IT application systems**

Establish application workloads and benchmark programs focusing on energy consumption and costs model. Reveal the relationship between IT value and energy consumption and costs, and describe it mathematically. We will draw on successful experiences of time complexity in algorithmic science and establish the complexity theory of energy consumption. We will also establish a value optimization methodology, subject to energy consumption and cost limitations. Make use of network effects to reduce energy consumption and cost. Study the low-cost and low energy consumption computer components and systems architecture; reusable and recyclable computer components; low energy consumption and high-efficiency virtualization, reconfigurable computing and hybrid computers with special/general-purpose components. We should explore information terminal products that never need to be recharged.

**5. Build up a nationwide e-environment platform**

Build up a nationwide environmental information network platform, environmental information resource sharing platform and environmental information service platform. Advance China's environmental information management towards goal of "digital environment protection", to keep a synchronous development along with the national informatization process. Help achieve balanced development of the economy, society, environment and resources.

# 8 Developing New Information Science and Interdisciplinary Sciences Based on Computation

The in-depth and extensive applications of IS&T bring forward substantial new challenging issues and also provide broad development space for information science and technology. The deep and extensive interaction among a number of different scientific fields has brought about a lot of common issues of research in different scientific and technological fields. It has also inspired new research ideas, methods and technical means that can influence each other. These characteristics in the development of S&T have facilitated the frequent appearance of interdisciplinary subjects. IS&T are facing new challenges and opportunities in development. It will be important to research new objectives for IS&T and other related fields in order to face these challenges, seize opportunities, develop new technologies, meet the ever-changing national demands and social needs, and promote the development of computation-based interdisciplinary sciences and the formation of new scientific and technological fields.

The development of modern computing and information systems, particularly the development of networked systems embodied by the Internet and distributed computing systems, has brought about many new characteristics of system behaviors. The increases of the system scale, the distributed features and the interactions among different agents in the system have produced unprecedented complexity. The theory of networked computing, Internet content retrieval, distributed and interactive algorithm design, new trustworthy software theory and techniques, and new computing principles and computation models will be important topics for basic research.

Exploring the nature of intelligence and understanding the human brain and its cognitive functions will be one of the most challenging undertakings in modern science. Theoretical and methodological breakthroughs of intelligent information processing based on cognitive mechanisms may enable breakthroughs in development of IS&T. The development of new intelligent S&T will be an important target for the next 50 years. Study of the nature of intelligence and the realization method is a cross-disciplinary subject of brain science, cognitive science, computer science and artificial intelligence. Intelligence science not only needs to conduct functional simulations, but also needs to study and explore new concepts, new theories and new methods based on cognitive mechanisms.

Significant progress in biology such as the completion of the complete genome sequencing of Saccharomyces cerevisiae, the most important model organism sequencing for the Caenorhabditis elegans as multi-cellular animals and the human genome working draft release, has not only promoted the development of biology, but also presented major research problems for information science. The complexity of the biological systems comports not only in the complexity of informational structures of DNA and proteins , but also in the laws of implementation and operation of information. With the continuous accumulation of relevant data on life sciences, the understanding of this complex process of the course of life will also enter a deeper level. Modeling and simulation of life system and understanding of the essential features and laws of life processes are common missions for scientists in the fields of computer science, life science as well as many other sciences.

Social computing has become a new international frontier for research and application following scientific, engineering, and biological computations. Based on cognitive science, intelligent science and science of complexity, research and application of social computation has become an urgent task to ensure national security, build up a harmonious society and implement the scientific development concept. The overall targets of social computing research are to form a set of new theories of computation, as well as to provide a set of advanced information processing techniques for a new information and knowledge society.

## 8.1 Target and Roadmap

Short-term targets (2010–2020): Research interactions of algorithms; determine the basic mechanisms of interaction. Break through key technologies of component-based software and software featuring high credibility. Research and develop computer systems with the ability to reason, perceive and learn, and make breakthroughs in the key technologies of brain signal processing and brain-machine systems. Achieve data integration of genomics, proteomics, structural genomics, structural proteomics, as well as various nomics. Construct parallel social systems in some typical application areas.

Mid-term targets (2021–2035): Establish a new theory of distributed and interactive algorithm characterized by a number of algorithms working together. Set up the design methodology for highly trustworthy software systems. Develop computer systems with thinking capability and multi-mode interactivity, and make breakthroughs in the key technologies for building machines with some mental power. Establish the theory of biological origin and dynamics of evolution based on system biology. Achieve experimental social computing and determine implementation and feedback mechanisms for decision-making for the management of complex systems.

Long-term targets (2036–2050): Modeling evolution and human intelligence, and develop a computational theory with the characteristics of evolution and human intelligence. Develop computers with abilities for emotion understanding and natural interaction, and develop brain-like computers. Realize simulation of living organisms and systems and their state prediction (including normal and disease states). Achieve widespread normal applications of the concepts and methodology of social computing (as shown in Figure 8-1).

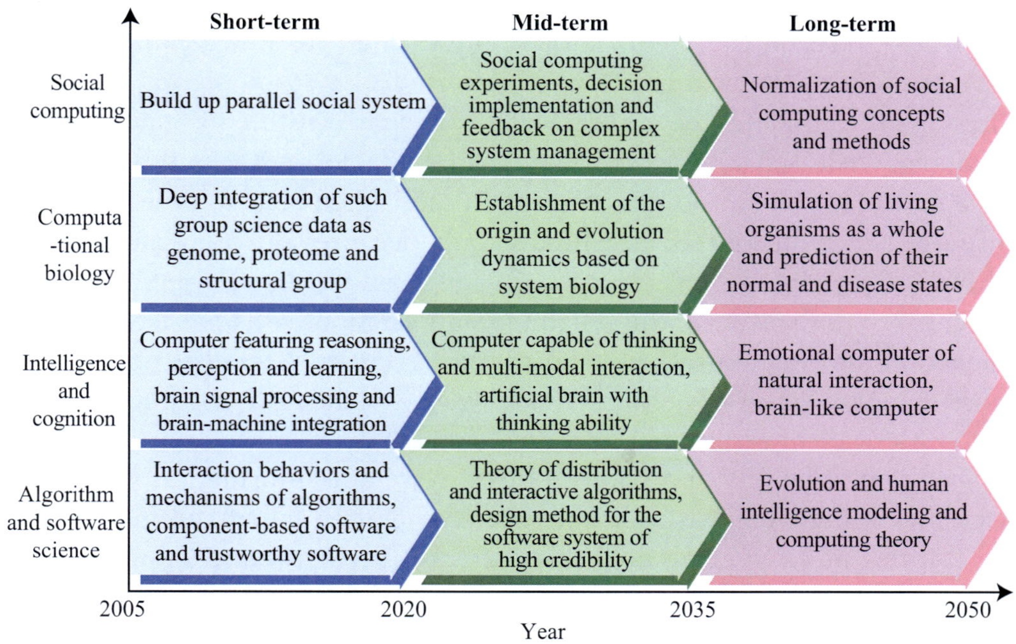


Figure 8-1 Roadmap for information science and interdisciplinary science

## 8.2 New Computation Models, Algorithm Theories, and Fundamental Theories of Software for Trustworthy Computing

### 8.2.1 Strategic requirements

Basic research on models of computation gave birth to a universal electronic digital computer in the 1930s–1940s. The popularization of computers and the rapid development of S&T in various fields have put forward increasing demands for computing. During this process, the Turing machine model is being challenged, and computer scientists begin to explore new computational models and theories in an attempt to break through the

Turing model's limitations. With computational thinking penetrating the human-machine-physical ternary world, the behaviors of computing systems demonstrate many new characteristics; in particular, the scalability, distribution, and interactions among multiple computing agents have allowed computing systems to show unprecedented complexity. Algorithmic research will focus on the interactions and collaborations among multiple algorithms rather than the design and analysis of single algorithms.

With increasing demand of applications, the social dependency on software will experience a steady rise. However, software systems have become larger and more complex, which has led to poor trustworthiness with low reliability and security. In 1996, the European Space Agency's Ariane rocket exploded 37 seconds after its launch due to the overflow generated when a 64-bit floating-point number was converted to a 16-bit signed integer in the program. In 2002, the United States National Institute of Standards and Technology released a report showing that the total losses due to software flaws were up to 59.5 billion dollars annually in the United States. Some important software is not outsourced to China mainly because software enterprises of China are not able to ensure the software's quality. As computer network systems rely on software, the reliability and safety of software have a direct impact on information security and network security. According to statistics, more than 90 percent of network attacks have exploited existing bugs in either system software or application software. In recent years, trustworthy computing focusing on reliability and security of software and information system has been increasingly emphasized. The security we often talk about includes the capabilities of preventing deliberate attacks (security) and improving the robustness of the system itself to avoid system failures (safety). Trustworthy computing is not just a project management issue but there are a lot of basic theoretical issues that still need to be resolved. The software basis for trusted computing has become a scientific issue that must be addressed in the next decades. Chapter 9 of this book is devoted specifically to security issues while this section only discusses software reliability and safety.

In addition, the development of biology, economics, sociology and other subjects has also raised new challenges for computer science. Changes in the computing system environment, scale, modes and fields of applications require that models and theories of computation in computer science have a significant breakthrough, and also a significant innovation in theory and design methodology of software.

### 8.2.2 Important S&T problems

**1. New principles and models for computation**

The computation performed by modern computers is essentially mechanical transformations with the formal symbols (0, 1), whose theoretical model can be attributed to the Turing machine. In fact, the inherent time and space complexity of the problems (tasks of the computation) make many important tasks intractable. In order to make breakthrough in such problems with combinatorial explosion, we are constantly seeking out new principles as well as models for computation. As a matter of fact, Turing himself once put forth the "choice machine" model which is more powerful than Turing machines; however, its super-Turing computing power is dependent on non-Turing computable inputs. The cellular automata without initial shape limitation presented by Von Neumann also present more expressive ability than Turing machines. Von Neumann also envisaged hybrid automata with integration of continuous and discrete mathematics. Whether the dreams of the pioneers of computer science can be realized in next decades will be a grand challenge for computer scientists.

The emergence of computer networks and their rapid development have not only changed our life and work styles, but also changed our understanding of the computation. Before the Internet era, academia was mainly concerned with sequential computing, which was seen as a function from input to output; computing without halting was considered meaningless because it did not produce any output. In the network environment, the computing agents (processes) can complete specified tasks in the process of constant interaction with the outside world. For such interactive and concurrent computing phenomena, the traditional "function"-based theory is no longer applicable. In order to understand concurrent computing, experts have proposed numerous theories, but so far no consensus has been reached. How to establish mathematical models for concurrent computation, and thus provide a solid theoretical foundation for the design and analysis of actual concurrent systems will be another major challenge for computer science in the next decades.

One idea put forward for a positive response to problems with combinatorial explosion does not regard the proof of whether a problem is an NP problem in the worst case as its goal, but uses the practical constraints in the human-machine-physical ternary world to seek out the critical point for phase transition, and provides a positive solution for a particular application environment. Another idea is to draw lessons from natural sciences and social sciences. Modeling the development process of cells from the perspective of computation not only helps to understand the basic problems of how a large number of genes and proteins in biological systems coordinate to control cell metabolism and DNA repair, etc., but also possesses great significance for communication protocol design, parallel computing models and mechanisms research in computing systems. Through analysis of information transformation processes of molecule level life activities, computing systems that are principally different from the silicon-based electronic computers are likely to be invented.

From a computer science perspective, the research of quantum physics and quantum computing may also help us to create a new model of computation. Understanding how brain cells are "automatically organized" to form a system with complex functions and intelligence, and how evolution is occurring at a biological molecular level, is not only of fundamental scientific significance for understanding the way that human intelligence is generated, but also of great guiding significance in building up information processing systems with higher levels of reasoning ability and intelligent property.

**2. Transforming algorithm design and analysis**

Algorithm analysis based on the computability and computational complexity theories is a core of contemporary computer science. The basic theories and techniques of algorithms (that is, through the precise sequence of basic steps, output can be obtained mechanically from the given input data) will be still widely used. However, with the popularization and deepening of computer and network applications, design and analysis of algorithms are experiencing transformation. For example, some concepts in traditional algorithms such as the assumptions of centralization, certain start/end, mechanic steps, exact result, may all need to be revisited to facilitate the studies of the human-cyber-physical ternary universe, in areas such as networked computation, evolutionary phenomena, algorithmic game theory, gene expressions, networked service composition, and the self-organization of the physical networks. As for the challenges of scalability, low energy consumption, security and usability in information systems over the next few decades, it is needed to investigate other complexity metrics indicators besides time and space complexities, such as energy complexity and effort complexity.

**3. Algorithmic network theory**

Future information systems are likely to be networked computing systems composed of interactive multiple resources and multiple computing agents. In such networked computing systems, it is frequently not an algorithm, but rather an algorithm network that can complete complex computing tasks through interaction and coordination of multiple algorithms. At present, the basic concept of the networked algorithms has not yet been established, not to mention the algorithm network theory. It is also an important research topic as far as what mechanisms can be designed for a distributed computing environment in order to allow the system to achieve the desired state of equilibrium. The design and analysis for algorithm networks in networked computing systems, particularly distributed mechanism design methodology, will become a new direction for computer science.

**4. The theoretical foundation and design methods of reliable software**

From a principled viewpoint, improving the reliability and safety of software means reducing errors and improving robustness. In general, software reliability is more difficult to be ensured than hardware reliability. Even for NASA software systems, reliability is still an order of magnitude lower than hardware reliability. Hardware fault-tolerance can be achieved through the use of redundant components, but this method is invalid in software. Software is a product of humans' intellectual activity; so, in essence, its production can only rely on humans. After 40 years of efforts, software engineering still lacks the scientific basis and strict measures to produce high-quality and highly-reliable products with acceptable costs. So far, the methods and theories proposed to verify the accuracy of the processes and ensure software reliability are still fragile and powerless in the face of large-scale practical systems. Credible measurements are the basis for credible computing, but they still lack measurement methods and theories for the dynamic credibility of software. To meet the requirements of software reliability, it is needed to research software development methods based on new principles. The ultimate resolution of software problems shall thus depend on breakthroughs in formal semantics and the combinatorics theory.

Over the past 20 years, volunteer communities and industrial alliances have adopted such schemes as open source code, web collaboration and user participation to develop a large number of software with high standard, ranging from system software, middleware, application software, network service systems and so on. Such software is not only of higher quality, but also comparatively safer, more reliable and with lower costs. These practices are not just relevant to business models, which may contain scientific principles and engineering know-how on software design, development and maintenance (for example, Apache and other World Wide Web software development has benefited from REST software architecture principles). It is needed to have a thorough study and to form a new type of software engineering theory based on these principles.

## 8.3 Making Breakthroughs in Intelligence and Cognitive Science

### 8.3.1 Strategic requirements of intelligence and cognitive sciences

Exploring the nature of intelligence and understanding the human brain and its cognitive functions is one of the most challenging basic scientific problems in the modern era. A person's cognitive system is the long-term evolutionary result of the biosphere. The human brain's processing and utilization abilities far exceed any of the existing computer information processing systems. We face three major challenges: Theoretical and practical issues of communications, computing, control, recognition, reasoning, judging and decision-making, in the age of information explosion; The reality of having not much room by simply relying on improving existing computer performance; The dual pressure from the lack of new computing models and ever increasing demands on IT. Close cooperation of information science and cognitive science has become a historical necessity in uncovering the nature of intelligence and developing machines and information processing systems with higher intelligence.

The in-depth study of human cognitive processes and intelligent information processing methods based on human cognitive mechanisms will have a far-reaching and significant impact on the all of the natural and technical sciences. Intelligent information processing based on breakthroughs in theory and methodology of cognitive mechanisms is likely to drive the breakthrough development of future IS&T. Combining brain science, cognitive science closely with artificial intelligence, strengthening the fundamental and original research in this interdisciplinary field will drive the development of key technologies of intelligent information processing in China's economy, society and even national security. It will provide a theoretical basis for the prevention and treatment of brain diseases and brain dysfunction and improve national health levels for us and contribute to major fundamental theories of brain science.

After more than 50 years of research, particularly over the past 30 years, AI has achieved great success; and to a certain extent, has enabled computers to have the abilities to "listen", "speak" and "read." But there is still a long way to reach the ideal goal for AI. For humans, even if your writing is colorful or your words are too vague, we can still make right recognition based on context, which shows that the recognition method in human brain is not "pattern matching." The human brain is much better at identifying and processing fuzzy information than computers. In addition, because sample libraries stored in computers have separate judging and identification functions, when the data size in sample libraries is very large, search matches will become powerless; while knowledge remembered by the human brain is one integrated mass to its determination organ, and its mode of identification is decision-making thinking to find and use knowledge. At present, the high performance of computers is in sharp contrast with their low intelligence problem solving, Internet applications in wide range and rapid development have brought an excellent opportunity for the development of artificial intelligence.

### 8.3.2 Scientific issues and key technologies of intelligence and cognitive sciences

Intelligence science is an interdisciplinary science based on brain science, cognitive science and artificial intelligence. Brain science studies the natural intelligence mechanism from the molecular level, cellular level and behavioral level, establishes brain models and uncovers the fundamental nature of human brains. Cognitive science studies the human brain's mental activity processes such as human perception, learning, memory, thinking and consciousness. Artificial intelligence research employs imitation, extension and expansion of human intelligence with artificial methods and techniques to achieve machine intelligence. Intelligence science not only conducts functional simulations, but also studies and explores new concepts, new theories and new methods of intelligence from different mechanisms. The study of intelligence not only requires reasoning from the highest level, but also needs to learn from the very bottom. Intelligence science takes a scientific approach to nature of intelligent and behavior of open systems using a comprehensive and integrated method. Research on intelligence science will establish the theoretical basis for the intelligence revolution, the knowledge revolution and the information revolution, and provide new concepts, new ideas and new modes for the development of intelligent systems. The main scientific issues and key technologies in this field are as follows:

**1. The physiological mechanism of machine learning and learning**

Learning is a basic cognitive activity and an accumulation process that combines experience and knowledge. It is also a process of improving system behaviors through the understanding of backward and forward linkages to external things. Neurobiological base of learning involves plasticity changes in link structures and synapses between nerve cells. This area has become one of the most active research fields in contemporary neuroscience.

Machine learning is an extremely active research area at present. This type of learning generally makes observations of the external environment to simulate human learning activities so that the machine can have access to new knowledge and new skills. This type of learning is also known as observational learning. Another form of study is introspective learning. Introspection refers to inspecting one's own thoughts or feelings; that is, self-observation. It also refers to the observation of one's own sensory and perceptual experiences carried out in controlled experimental conditions. The introspective method is a kind of research method from the early days of psychology that studies and processes psychological phenomena according to the examinee's report or described experiences. Introspective learning is a method of learning used to introduce the concept of introspection into machine learning; that is, through inspecting and caring for the knowledge processing and reasoning method of the intelligent system itself, uncovering problems from failures or inefficiency and forming learning goals for self-repair, thus improving ways of problem-solving themselves.

**2. Perceptual information processing**

The representation of perceptual information and the organization and integration of overall perception are basic problems of perception study, in addition to the foundation of studying other cognitive processes at all levels. Research in this area should form characteristics of the relationship between the local nature and the large-scale nature of perception, establishing a systematic theory of perception. Research should also combine the experimental study of cognitive neuroscience and computer perception research in computing theoretical levels, the brain levels of knowledge representation and computer implementation levels, and propose brand new theories and solutions.

**3. Language cognition**

The development of psychological science has revealed that language acquisition, understanding and generation are extremely complex issues. The study of language provides us with a unique perspective to understand the interactions between descendible heredity and the environment. The research of language not only enables us to have an in-depth understanding of environmental and genetic effects in language acquisition, understanding and generation, but also provides important revelations for us to further understand and learn the role of the gene in other varieties of cognitive functions, and gives us a good opportunity to reveal a variety of cognitive and intellectual processes of human nature. Current language cognition mainly studies natural language processing, particularly Chinese information processing.

**4. Memory**

The aim of memory is to produce an accurate representation of past experiences and extract and use them correctly and efficiently. Therefore, encoding and retrieval must experience top-down control. In the encoding stage, there is a need to integrate multiple disparate features or combine several blocks of knowledge in order to form a unified characterization and separate the similar events. Focusing, monitoring and verification must be carried out during extraction in order to efficiently select relevant information and suppress irrelevant information. The control of memory processes and the associated neural mechanisms are quickly becoming hot scientific issues. An in-depth study of the neural physiological mechanisms of memory can improve the understanding of humans' intelligence activities, and can also provide a theoretical basis or reference for the establishment of effective intelligent information processing systems.

Working memory is a part of short-term memory, and its function is to preserve information for a short period of time for processing. It is the core link of a variety of higher cognitive functions closely linked with human consciousness and intelligence. Studying the structure and function of working memory is of great significance in understanding the nature of human intelligence. Different memory systems adopt different organization methods. For instance, input channels are a key factor in short-term memory organization; long-term knowledge memory relies on hierarchical semantic networks; and the relationship between things is the basis of episodic memory organization.

**5. Thought**

Thought is a cognitive representation in response to certain aspects of the external world (including humans themselves) and the handling of these representations and understanding. The information representation and processing period is related to concepts and reasoning as well as problem solving and other higher cognitive processes. The study of thought is of important scientific significance and valuable to understand the nature of human intelligence and develop artificial intelligence. Studying different levels of thought models as well as rules and methods of thinking will provide principles and models for new intelligent information processing systems.

**6. Affective computing**

Affective computing is related to emotions, feelings generation and emotional factor influence that aims to give computers the ability to identify, understand, express and adapt to humans' emotions. Its aim is to allow the computer experience emotions; that is, enable the computer to possess intelligence. Affective computing is one of the foundations in building a harmonious human-machine environment. Wearable computers and post-PC computers will provide affective computing with convenient research platforms and manifestations; affective computing will have great application value and potential for development. Contemporary cognitive scientists have compared the feeling to classic cognitive processes such as perception, learning, memory and language, and the research of the emotion itself and interaction between emotions with other cognitive processes has become a research hotspot of contemporary cognitive science.

**7. Consciousness**

The origin and nature of consciousness is one of the most important scientific issues. It is very difficult to give consciousness a united and exact scientific definition at present. The understanding of sense is changing in different areas. Nobel Prize winner Crick believes that consciousness involves the neural mechanisms of a combination of attention and short-term memory. They can be studied using scientific methods. Crick's amazing assumptions on sense and specific recommendations in the study of visual awareness through visual attention and short-term memory have aroused a wide range of interest from a large number of cognitive psychologists, neuroscientists, and computational neuroscience scientists.

Consciousness is related to the higher cognitive processes, such as perception, attention, memory, representation, thought and language, of which awareness is the core. In recent years, the development of cognitive science, neuroscience and computer science, particularly new non-invasive experimental techniques, have enabled the study of consciousness to become a common study hotspot for a number of disciplines. In order to uncover the laws of sense, and build up the brain model of awareness, it is needed not only to study consciousness cognitive processes, but also to research unconsciousness cognitive processes, namely, the automatic processing of information in the brain as well as the mutual transformation mechanisms of the two processes in the brain. It is also needed to study the awareness-cognition theory, neurobiological theory of consciousness and the information processing mechanism of the consciousness and unconsciousness.

## 8.4 Promoting Computational Biology as a Main Branch of Life Science

In 1996, the complete genome sequence of Saccharomyces cerevisiae was completed, and it became the first eukaryotes sequenced, which was a landmark in the history of genetics and molecular biology. In 1998, Caenorhabditis elegans, the most important model organism of multi-cellular animals, was sequenced. In 2003, the working draft for the human genome was released and a number of media hailed it as a revolutionary achievement. At present, hundreds genome sequence of species have been completed, which has provided scientists in many fields with an unprecedented amount of information. For the analysis and interpretation of this data, bioinformatics has become crucial to genome research and has made steady strides in development. In the present and future decades, computational biology with the core of bioinformatics will continue its in-depth development in the following areas:

**1. Information analysis on splicing, assembly and gene annotation, etc, in large-scale genome sequencing**

A large-scale sequencing and analysis of information will be the most basic task of genome research. Data on the human genome, rice genome as well as other complete genomes are all obtained through such means. In the coming decades, the genome sequencing of different species and races will become routine. The human genome has three billion base pairs, and considering 10-fold sequencing coverage, it will generate at least 30G of raw data, which demands the high performance activity of stitching and assembling algorithms. Coupled with a large percentage of repeat sequences, the difficulty of genome splicing and assembly will be greatly increased. With the emergence of such high-pass sequencing platforms as Solixa, Solid and 454, a vast amount of data will be generated, which will put forward higher demands for splicing and assembly algorithms, so we must consider issue of time complexity of algorithms, and also solve the bottleneck of memory in the process of splicing. Only by developing the corresponding algorithms and software can we take the advantages of high sequencing depth and obtain a more refined genome spectrum. Meanwhile, with the large-scale generation of transcript data, the workload of gene annotations will be greatly increased, especially the method of using comparative genomics to have genes annotated quickly to completely new genomes that are newly measured. With the increasing emergence of whole-genome data, in order to complete a comprehensive genome annotation, it is needed to compare the genomic and transcript data of dozens, hundreds or even thousands of neighboring species, which needs more effective methods for gene annotations.

**2. Discovering and identifying new genes and new SNPs using EST data and genomic data**

EST sequences (Expressed Sequence Tags) are short cDNA sequences of gene expression, and they carry certain fragments of information of the complete genome. Presently, number of human EST sequences is more than 10 million in GenBank's EST database (dbEST), which covers about 90 percent of the human genome or more. All of the data on EST is obtained from experiments and it is a valuable resource to study transcriptomes and genomes; so using such information to discover new coding and non-coding genes and single nucleotide polymorphism (SNP) through large-scale computing has become an important research topic in recent years and will be more important in the coming years. Due to the large amount of EST data coupled with the development of biotechnology in recent years, the speed of producing EST data is experiencing rapid development, and mining new genetic information and SNP information from these vast amounts of EST data requires generating new algorithms and software. At the same time, EST data and genomic data will often be analyzed together, so the integration of the vast amount of EST data and vast amount of genomic data has put forward high requirements for algorithms and storage.

**3. The research on the origin of genetic codes and the evolution of biology**

Since the release of Darwin's *Origin of Species* in 1859, the theory of evolution has become one of the most significant contributions to the development of natural science and natural philosophy. The core of the theory of evolution is describing the history of biological evolution (phylogenetic evolutionary trees) and exploring the mechanisms of evolution. Since the middle of the last century, along with the continuous development of molecular biology, evolutionary studies have also entered the molecular level. Molecular evolution research based on nucleic acids, protein sequences and complete genomic information will be an important research topic over the next several years and will provide a new basis for a natural philosophy of understanding the origin of life and evolution. The work of comparing two or more complete genomes requires new ideas and methodologies and needs new, efficient algorithms and software; hence the information of evolution can be extracted from a large number of species genomes and transcriptomes. The construction of phylogenetic trees is based on the gene or sequence differences, and after taking complete genome information into account, the description of these sequence differences has become extremely complex and at the same time has also presented high storage requirements. Information on evolution is hidden among the differences of genomes or the transcriptome, so identifying and measuring these differences fast and effectively requires the support of new algorithms and software. Evolution analysis based on complete genomes or transcriptomes will promote the study of biological evolution.

**4. Analysis of the gene function expression spectrum based on large-scale micro-array data and proteomics data**

After large-scale genome sequencing, in order to understand how genes function, it is needed to understand the genetic functionality expression spectrum. According to the viewpoints of many mathematicians and physicists, the static genetic map present in the human genome should be spread to the time and space dimension. In order to obtain the function spectrum of gene expression, new technology has been developed on two levels of nucleic acids and proteins. This technology includes DNA chip technology at the nucleic acid level and large-scale protein separation and sequence identification techniques in protein levels, also known as protein spectrum technology and proteomic research. The development of these technologies has not only brought with it a large amount of data, but also presented further demands in terms of new computing theories and methods. At present there is a large amount of chip data on the same platform, and also an enormous amount data on a large number of different platforms, so how to best integrate this massive experimental data onto the same platform and other various platforms will be a fundamental problem for large-scale DNA micro-array analysts in the future. Meanwhile, with the rapid development of bio-chip technology, more platforms and chip data will be produced, which will require more effective algorithms and software to handle the data on the same platform or different platforms. Gene expression spectral data and protein spectral data are more than simply one-dimensional digital data; they not only contain image data, but also data expanded from multidimensional levels of time and space, and processing such data creates a high demand on the computer's memory.

**5. Protein structure simulation, molecular design, and medicine design based biological macromolecule structure**

The problem of protein folding; that is, how the nascent polypeptide chain is folded and assembled into a state with biological activity in the cell is a fundamental problem needing to be answered to ultimately clarify the central dogma and an important issue for biological physics in the 21$^{st}$ century. The results will provide the basis for the modification of natural biological macromolecules and medicine design based on the receptor structure with an obvious application foreground. One of the objectives of Human Genome Project is to elucidate humans' 100,000 kinds of protein structures, functions, and relationships with a variety of human diseases, and search for treatment and prevention methods for a variety of diseases, including medicine treatment. Medicine design based on biological macromolecular structure is an extremely important research field in bioinformatics. In order to suppress the activity of certain enzymes or proteins, we can use molecular docking algorithms to design inhibitor molecules on the computer as drug candidates, on the basis of knowing their tertiary structure. This method of discovering new medicines has strong potential and great economic benefits. However, the current system of protein structure estimation and modeling is still far away from achieving the desired result and there is much room for development.

### 6. Bio-macromolecule interaction networks, information analysis and systems biology research on metabolism and disease arising pathways

Learning about real life processes requires an understanding of gene networks such as regulation pathways and metabolic pathways of bio-molecular reactions. By coupling the relationships among all the units and hierarchical levels together, the behaviors of biological systems can be simulated to predict how the system will evolve if it has been stimulated and interfered in by the outside world. Systems biology is thus generated based on these kinds of concepts. Systems biology affirms that: complex systems set up through different levels of associations are not simple superposition of the system. Some emergent behaviors and emergent natures of the law will be present in this complex system, and gradually show the structure and functional characteristics of organisms at the cellular or sub-cellular level; the living body is a complex and dynamic network, so research for life needs to start from the whole to study these complex interactions and regulatory relationships in the network of life, not just study isolated nodes in the network. The interaction network of biological macromolecules is a complex multi-level and multi-element network, and there exists a longer side consisting of different types of interactions or regulation relationships in this network. There are also some nodes consisting of molecules of different natures. At present, there is no ready-made method that can analyze the nature of such a network, which presents new challenges to network analysis algorithm. Through network analysis, we can gain an in-depth understanding of metabolism and the pathways of disease occurrence. Explaining pathways of diseases from the perspective of networks and systems biology requires the integration of information at various levels and needs new ideas and algorithms.

### 7. Research on the non-coding sequence, non-coding genes and non-coding RNA in the genome

With the completion of the human genome project and continuous accumulation of the mammalian transcriptome data, we can see that only a tiny fraction encode protein in the genetic material of the human and other senior eukaryotes, or about 3 percent of the human genome. Hence we are still unclear about the functions of the remaining 97 percent of the DNA sequence, which is habitually labeled "non-coding DNA." In recent years, a large number of new results from experiments have shown that non-coding DNA can be expressed, and the expressed product is a dynamic information carrier for life processes, known as non-coding RNA (NcRNA). Non-coding RNA is a major component group of transcription and its complex and diverse functions greatly expand our awareness in terms of the complexity of gene transcription regulations. With a large number new non-coding RNA discovered in recent years, research on biological functions of non-coding RNA has received unprecedented attention. Non-coding RNA may play an important role in the process of proteins secreting to a wide range of gene regulatory cells. Since 2000, significant progresses in this area were repeatedly included in the top ten scientific discoveries of that respective year. Research on non-coding RNA and its gene has become a hotspot in life sciences in recent years and this trend will continue in the coming decades.

The emergence of non-coding genes has revived scientists' interest in the "RNA world" and the interest in the proposition of "the origin of life from RNA molecules." Some scientists believe that, tens of thousands of non-coding RNA molecules of protein molecules can form a huge molecule network and regulate life activities in cells and correspond with protein—the protein interaction network is like dark matter in cosmology and will provide an unparalleled bright future to genomics and life science research.

Bioinformatics has developed rapidly not only in basic research but also has broad application prospects including:

(1) The development of databases, database tools and network tools, which can effectively support the growing research needs of bioinformatics.

(2) The development and collection of disease-related genetic information and related algorithms and software.

(3) The construction of a genome database related to plant and animal breeding and the development of molecular marker-assisted breeding technology.

(4) Research and development on medicine design software, and bio-informatics-based molecular biology techniques.

Bioinformatics can not only promote the development of life sciences, but also may bring inspiration in the fields of mathematics and physics, such as DNA computing and DNA computers, a new computing architecture inspired by bio-networks.

By 2020, the cost of human genome sequencing can be reduced to 1000 dollars/person, which will make it possible to detect individual genetic composition. It is estimated that by 2020, the prediction accuracy of protein spatial structure can move closer to the experimental observation level.

By 2035, super-databases reflecting the main differences of the human genome (HUMAN GENETIC DIVERSITY) will be able to be built. Current research shows that the differences between two separate individual genomes are about 0.1 percent. In other words, with $3\times10^6$ base pairs and the world population at 6 billion ($6\times10^9$), the total base differences between human groups are $18\times10^{15}$, storing and computing such data can be achieved through computing technology. If taking the cost of sequencing into account, it is also possible to measure the personal data of one million and 10 million levels. This will provide the most important basic data for human health.

By 2050, we will have a deeper understanding of the complexity of biological systems. Life is not a clustered unorganized group of protein molecules, but rather is highly organized. Biological molecules form cells, cells form structures, and structures form organ systems and these thus constitute a system. Nucleus and cytoplasm are mutually affected; there are connections between cells; different organs may have different functions; and there exists synergy between organs. Therefore, a normally existing individual organism is extremely organized, multi-leveled and dynamic. It took millions of years to complete evolution to go from a disorderly process to an orderly process and go from the simple to the complex. Therefore, the complexity of the organism is not only reflected in DNA, the complexity of the structure on protein and other information, but also in the implementation and operational law of the information itself. With the continuous accumulation of relevant scientific data in life science, our understanding of this complex process will also deepen.

## 8.5 Promoting Social Computing as an Important Field of Social Sciences

### 8.5.1 Strategic requirements of social computing

Social computing is an emerging field of research and development in computational theories and methods for social activities, processes, organizations, functions, and applications. Based on a combination of cognitive science, intelligence science and complexity science, social computing research and applications have become an urgent agenda for ensuring national security and for building a harmonious society. The current needs and challenges of social computing are as follows:

**1. Social behaviors and emerging social patterns**

The rapid development of new Internet technologies has made the entire community realize that these new technologies will profoundly affect the structure, organization and activity patterns of the future society. Although we are unable to know exactly what form the society will take on in the future, yet it is still clear that, as opposed to traditional social patterns, its dynamic changes are faster and more difficult to predict, and its organization forms are even broader and more unfathomable. These factors have made social computing a new and important area for further research.

With the advances of science and technology, especially those in network communication technology, the social networking trend has become increasingly evident. The network includes not only physical communication networks and the Internet, but also social networks emphasizing human-to-human interaction. Facing these highly networked realities, we need new computational theories and tools to understand everything around us. The social computing meets people's needs in this aspect.

**2. Scientific analysis and decision support of complex problems**

China is on the historical phase of fully achieving a well-off society, promoting the process of modernization and building a socialist harmonious society. The spectrum of the social development sequence corresponds exactly to the "non-steady-state" development phase. In response to a series of major issues concerning national economy and people's livelihood, especially in the face of natural disasters and social emergencies, we urgently need a scientific method which provides strong support to the analysis and decision-making in these situations. Research and applications of social computing will provide such a solution to this quandary.

**3. Knowledge-based industrial engines and enabling technology**

For enterprises in the highly information-oriented and competition-oriented society today, how to obtain accurate information is not only a driving force for development, but also a key factor in a company's survival. Knowledge-based industrial engines provided by social computing provide enterprises with effective information retrieval tools, and the popular enabling technology offers a highly efficient operation system that facilitates the operations and managements of those enterprises. All these technologies have become the driving force for the enterprises' rapid development.

### 8.5.2 Scientific issues and key technologies of social computing

The overall goals of social computing research are to form an advanced new computational theory based on the information and knowledge society, to develop advanced information processing technologies to deal with social problems and to realize universal applications. To accomplish the above goals, following five major tasks need to be completed in succession:

(1) Real-time access to comprehensive information. Computing, analysis and processing are based on the access to information. Surrounded by diverse social information, how to carry out effective real-time information collection is the most urgent task at present.

(2) Content computing and understanding. After having collected the information, it is needed to calculate and understand the content, extract the required data and provide data support for further modeling and simulation.

(3) Build up parallel social systems. Simultaneously emphasize artificial systems and actual system. In other words, compose parallel social systems, with the simultaneous running of the two systems focusing on the constitution of parallel execution. The main purposes of constructing parallel social systems are to compare and analyze the behaviors of the actual and artificial systems based on their mutual connections, complete the "study" and "estimates" of their respective future states, correspondingly adjust their respective management and control system and implement effective solutions to complex social problems.

(4) Computational social experiments and decision support. In the computational experiment method, traditional computer simulation has become a "trial" process in the "computational laboratory", as well as a way to "foster" all kinds of complex systems. However, the actual system is only one possible outcome of the computational experiments. Through parallelism and

the reference of artificial and actual events in the parallel system, a particular computational social experiment can be carried out to achieve effective control and management of the actual system, so all relevant actions and policies can be experimented on and evaluated.

(5) Decision-making execution and feedback system of complex system management. Use the parallel system approach to manage and control the complex system. The parallel systems change the conventional passive simulation to an active parallel artificial systems, also change its role from passive to active, static to dynamic, offline to online, and finally from a subordinate position to an equal status. By doing so, the artificial system fully plays its role in management and control of the actual system. It also enhances the feedback from the actual system on decision-making information, achieves automatic adjustments and controls, and increases the accuracy of the computing social experiments as well.

The main research areas of social computing are to design, implement, evaluate and promote a variety of information technologies of people-to-people communications, coordination and cooperation in society on the basis of advanced computing technology, focusing on people and activities as well as interdisciplinary collaborative research. The major scientific issues and key technologies of social computing can be divided into the following three layers:

(1) The technology layer, which includes some basic techniques supporting social computing. Technologies currently involved include data acquisition and organization technology, machine learning and data description, content understanding and opinion mining, Web science, dynamic visualization techniques, computational psychology and computational sociology, etc.

(2) The model layer, which contains those methods of model construction for specific social issues and social phenomena supported by basic technologies. It mainly includes cognition and behavior modeling, uncertainty in complex networks and complex systems, comprehensive integration and human-computer seminar, and so on.

(3) The decision-making layer, which means artificial society and parallel management. Built on a wide range of complex models, it can strongly support the final decision by providing accurate simulations and simulation results.

We have divided the research of and development strategy for social computing into four phases. Each phase has its corresponding focus and research contents, and is comprised of the desired research results and specific applications.

The goal of the first phase is to realize typical applications in basic social computing theory and major issues, and to reach an advanced international level. Typical applications include social monitoring and public opinion analysis regularly or irregularly, especially the monitoring and coverage for emergencies; advanced intelligent control of urban transportation and industrial chemical processes; research of e-commerce mechanisms and e-commerce computational experiments. The expected theoretical result of this phase is a system of socio-physical methods.

The objective of the second phase is to improve social computing theory and technological framework, to achieve the wide range of applications in important areas and to form major decision-making platforms and software with completely independent intellectual property rights. On the basis of the first-stage applications, this phase has been expanded and consolidated, with the main applications including: based on public opinion monitoring, build an integrated social information system and improve the emergency management mechanisms for social security affairs; based on the intelligent control of urban transportation, implement intelligent management controls for other public facilities and the environment; extend industrial chemical process controls to the area of complex production process control; based on e-commerce mechanisms research and computational experiments, build an intelligence analysis and decision-making system for business. Theoretical research of this stage will focus on computational psychology, which will guide specific applications.

The goal of the third phase is to promote more scientific government and enterprise decision-making ability and implement the concept of scientific development based on wider applications of social computing. Focusing on the further integration of the former system-level applications, this phase has achieved the digitization of various social functions including digital government, digital economy and digital enterprise, etc. At the same time, social computing has become an integrated discipline featuring theoretical frameworks and technical supports based on continuous improvement.

In the fourth phase, based on the promotion of a wide range of applications and the popularization of social computing in education, normalization and socialization of social computing concepts and methods will be realized. We will have expanded and integrated social functions of the aforementioned digital applications, eventually resulting in a complete digital society and constituting an engine of knowledgeable economy. At the same time, on the basis of the discipline of social computing, we will have developed social macro-informatics; and thus far, our understanding of society will have reached a new level.

# 9 Establishing a Technical System for National and Social Information Security

## 9.1 Target and Roadmap

Short-term targets (2010–2020): Construct a network-based lightweight information security foundation technological system. Research the establishment of a large-scale and collaborative cyberspace security technological system. Research and develop an integrated security service technological system. Achieve the commercialization of point-to-point optical fiber quantum cryptography communication systems and the trial success of a 70km metro optical network quantum key distribution.

Mid-term targets (2021–2035): Build the base technological system for automated and intelligent information security. Research the establishment of a highly credible and interactive cyberspace security technological system. Research and develop the technological system for automated information security services. Achieve the commercialization of metro quantum cryptography communication systems, and expand Metropolitan Area Networks to Inter-metropolitan Networks.

Long-term targets (2036–2050): Build up a dynamic and high-intensity-based information security technological system. Research the establishment of high availability and self-organized cyberspace security technological systems. Research and develop intelligent security service technological systems. Implement global practical secure communication networks based on quantum key distribution (as shown in Figure 9-1).

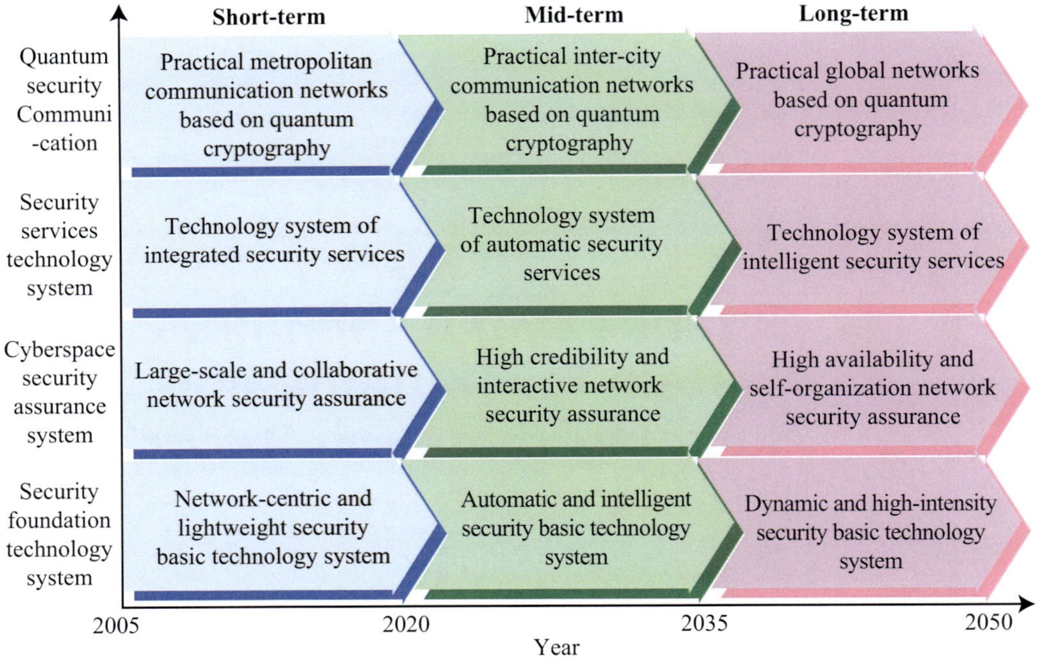


Figure 9-1 Roadmap for information security technology

## 9.2 Establishing a Basic Technical System for Information Security Based on Cryptographic Techniques

### 9.2.1 Strategic requirements

The in-depth development and extensive application of IT, particularly growing computing power and fast-changing computing models, has brought great challenges to the existing system of information security technology. We urgently need to establish an information security technical system with a higher level of security that can meet the needs of a variety of applications.

Cryptographic technique is one of the important methods to protect information security and the core base of the information security technological system. The development of cryptography will be the key to establishing an information security basic technological system.

The security involved in existing information security technologies mainly refers to computing security, and with changing computing power and computing models, this kind of security is either reduced or disappears entirely; therefore it seems particularly important to research information security technology with unconditional security (that is, having nothing to do with the computing power). In addition, with changing application modes, it is also very important to research information security technology that can adapt to different environments and modes.

### 9.2.2 The main scientific issues and key technologies

**1. High-intensity cryptosystem design and analysis**

The design and analysis of high-strength cryptography is not only a technical or application issue, but also a complex theoretical issue directly associated with mathematical theory.

**2. Dynamic security mechanism design and analysis**

To meet the demand of dynamic and complex application environments, we must research the establishment of dynamic security mechanisms, whose design and analysis are both complex and difficult.

**3. Unconditional security mechanism design and analysis**

With growing computing power and fast-changing computing models, we must research the establishment of security mechanisms that have nothing to do with computing power (aka unconditional security). However, problems in design and analysis will be a major challenge.

**4. Establishing an axiomatic information security technological system**

Many problems in the existing information security technological system can not be quantified or axiomatic and are unable to be accurately characterized and quantitatively analyzed; hence they require the development of an axiomatic information security technological system to solve related scientific problems.

**5. The key technologies for automation and intelligent information security**

As a practical technology, it is necessary that the degree of automation and intelligent level in information security technology increases to meet the rapidly growing technology needs and a great variety of application requirements; hence we must make breakthroughs in a number of critical technologies on automation and intelligent information security.

**6. Information security problems caused by the development of new technology and applications**

A lot of important information security issues will inevitably emerge in new technology development and the application process, so for the development of new technologies and applications, it is necessary for us to do advanced research on one hand, and on the other hand solve security issues in application in a timely fashion. This is at once both practical and challenging.

## 9.3 Establishing a Technical System for Cyberspace Security Based on Supervision Technology

### 9.3.1 Strategic requirements

Monitoring supervision and management techniques are the key to building cyberspace security technological systems, but with the continual emergence of new technologies as well as ever changing application modes, monitoring supervision and management techniques are facing great challenges, and it has been difficult for existing technology to cope with ever-changing application requirements. Therefore we need to follow up in real time by conducting innovative research.

The integration of space and ground brings with it more and more extensive coverage of information systems, which will present new demands for the safety and security of cyberspace. Meanwhile, with the increasingly high degree of social information, changes in the size of information systems and degrees of complexity will put forward higher requirements for the security of cyberspace.

It can be predicted that, with the in-depth development and extensive application of networks, web attack techniques will constantly evolve, and their infiltration capacity, viable capability and self-organization capability, etc., will continue to be improved, which in turn will put forward more serious challenges to cyberspace in terms of safety and security.

### 9.3.2 The main scientific issues and key technologies

**1. The purification technology on cyberspace**

In order to create a green and healthy cyberspace that promotes a cultural environment, we need to research content purification technology for backbone networks and mass storage space, establish the mechanism of high-speed content identification, verification, classification, filtering and tracing, research harmless web content conversion technology and achieve full-effect cyberspace purification.

**2. Active immunization technology on security threats**

For the establishment of safe and reliable operation platforms for cyberspace spaces, we need to research active immunization technology to combat against security threats, in order to achieve the characteristics of self-identification, isolation, rejection and confrontation of security threats on the infrastructure level with automatic deployment and upgrading capacity of immunization resources; thus forming an initiative, depth-based and progressive-oriented cyberspace security system.

**3. Security technology's interaction with the social credit system**

In order to build up effective constraint mechanisms for a diversified network space, we need to research security technology as it interacts with the social credit system, use the conduct audits, determinations and documentation based on password technology, form a social resultant force against security threats aiming at misconduct in cyberspace and promote cyberspace purification from the social constraint level.

**4. Effective monitoring control core technology and the corresponding coping strategies**

With the continual emergence of new technologies coupled with ever changing application patterns, a great challenge has come in terms of monitoring regulations. Therefore, we should not only impel the in-depth research of efficient monitoring supervision and management techniques, but also explore new strategies for coping as well as other technological solutions.

**5. Support unified certificate authority audit techniques for multi-network integration**

Along with the development of multi-network integration, business systems require a separation between the business itself and such security features as authentication, authorization and auditing, the support of multi-factor access authentication and interoperability, the support of cross-domain authentication, privacy protection and the implementation of the access and control of dynamically complex fine-grained functions. Such requirements have brought about huge technical challenges.

**6. The infrastructure component technology of network information security**

The information security infrastructure component is a key component for security systems to build up the information security infrastructure and application security support systems. The design and development of security systems is bound to enter into a highly-efficient, highly-reliable and intelligent stage, so we need to deepen our research and further develop network information security basic component technology.

**7. Discovery and early-warning technologies for unknown malicious codes**

The development of malicious code technology has posed enormous challenges to traditional means of protection, and the protective effects of single-point protection or the centralized protection model are becoming increasingly limited. Therefore, it is particularly important and critical for us to do further research on issues such as the identification, analysis and feature extraction of unknown malicious codes.

**8. The collaboration protection technology of large-scale network systems**

In order to improve detection capabilities for malicious codes and protection levels for various protection nodes in large-scale networks and build up an advanced collaborative platform for malicious codes, we must effectively resolve certain key technical issues such as correlation analysis detection of malicious codes, automatic generation of protection strategy, and provide a coordinated response and full protection.

**9. The security storage and access technology of mass information**

Super large data scales coupled with special modes of data organization, etc. in the mass storage system have created special security requirements. For example, it is necessary to research such core technologies as breakthroughs in confidentiality protection mechanisms for massive information, breakthroughs in intrusion tolerance mechanisms and breakthroughs in integrity protection mechanisms.

**10. Build-up technology for trusted computing environments**

Trusted computing thinking will make a great impact on future information security systems. We need to strengthen the research on new trusted computing platform security architecture; the technologies of constructing an open trusted platform system and feature applications for trusted computing platforms based on the hardware level.

## 9.4 Establishing a Technical System for Information Security Service Based on Assessment Technology

### 9.4.1 The strategic requirements

With the extensive application of IT, services have become an important guarantee for the sustainable development of IT and the information security. The information security service has utterly dominated the information security industry and the security service has become the core link for ensuring that security technology and safety management form an integrated ability of protection. Therefore, the development of a technological system for information security services has become even more vital and pressing.

As a key link in the process of information system security engineering, information security evaluation occupies an important position throughout the life cycle of the information system. Its modes of measurement have changed from traditional security products, communication systems, operating systems and network systems to complete information security systems covering both technology and management. Information security test evaluation and information security services (such as security consulting, system planning, safe management and emergency response) will be gradually integrated to form a closed-loop security system of the information system life cycle.

From a development point of view, the security evaluation problem of software and hardware information products is one of the key issues that needs to be solved in a timely manner. The information system has become increasingly complex and its difficulty in assessing security will gradually increase. The rapid growth of networks will also raise serious challenges to the network security situation assessment.

### 9.4.2 The main scientific issues and key technologies

**1. Test evaluation and service technology based on benchmarks in information security**

Security benchmarks are the basis in achieving a standardized management system for safety and security, so we should pay close attention to the establishment of a unified database of information security benchmarks, research on benchmark-oriented automated test technology and service supporting technologies and the formation of a normative evaluation system and service capabilities for products and information systems.

**2. The integration of technological systems, management systems and service systems**

Comprehensive information security capabilities need the full integration of technology, management and services; hence we should focus on the integration of information security systems based on security benchmarks and establish an operational mechanism to coordinate security technology, management and service and develop an intelligent security system with capabilities of auto-building, auto-deployment and auto-updates of security resources.

**3. In-depth reverse test technology**

By means of reverse tests, we can further extract security flaws and risks existing in products or information systems, coordinate with positive tests and other assessment means and make comprehensive assessments and analysis of the security situation of the target in question. We should focus on researching in-depth reverse test techniques based on core technology products, looking at critical infrastructure and key information systems, forming automatic and intelligent measurement tools and supporting the in-depth development of a safety evaluation system.

**4. Security testing and evaluation technology of information products**

At present, security analysis of hardware and software products still lacks proper technical support. Key issues to be addressed include security quantitative measurement systems of hardware and software products, vulnerability identification and determination of hardware and software security products and the security quantitative analysis for the password, security protocols and other special modules.

**5. Security assessment methods and techniques for complex information systems**

With the full implementation of the related security policies and systems such as grade protection in China, one of the most important missions to be carried out involves a comprehensive security assessment of the information system. However, this assessment still lacks effective technical means to be effectively carried out at present. With the development of IT, this issue will become more and more prominent. The security assessment of complex information systems will be one of the core problems that need to be solved. This assessment will include security element extraction in the complex information system, an analysis of potential paths of attack in the complex information system, security measures in the complex information system and so on.

**6. Security situation assessment techniques of large-scale networks**

The development of attack technology has brought great challenges to existing security situation assessment technology. We need to address a number of key issues including but not limited to the non-destructive extraction of situational data, analysis and forecast of the security situation, detection and early warning of the new large-scale network attack and automatic generation and dissemination of security policy.

## 9.5 Establishing a New Network System with Communication Security Based on Quantum Cryptography

### 9.5.1 The strategic requirements of secure quantum communications

With the rapid development of the current online world, business circles and governmental departments require secure communications now more than ever. Existing information security technologies are facing serious challenges by unexpected progresses made in mathematics, high-performance computing and possible quantum computing. Therefore, people are starved for a new type of secure communication method to meet with the growing complexity of networks that can take into account users' different security requirements. Quantum cryptography is expected to be the next generation of secure communication technologies that will adapt to future needs.

The target in this roadmap is to achieve a new set of powerful quantum cryptography security systems to meet future demand for secure communication networks. This system should be characterized by convenience, easiness of application and long-term stability. To achieve this objective, we must promote the research of relevant basic theories and experimental physics, while at the same time developing many new components such as the quantum light sources, single-photon detectors, quantum repeaters and "quantum-friendly" network components, etc.

The core technology of quantum cryptography is quantum key distribution (QKD), which adopts quantum states to convey classical information (i.e. key). Because QKD is based on quantum mechanics, any eavesdropping behavior will be discovered by legitimate users. When legitimate users establish a symmetric key through cryptographic protocol and confirm there is no eavesdropping, they can use the security key provided to achieve secure communications. Combined with the "one-time pad coding" feature, non-decipherable secure communications that cannot be tapped will be provided. In the past ten or more years, quantum cryptography research has achieved many breakthroughs and a lot of key technologies are beginning to mature. At present, the technology is in the engineering research stage of practical applications and it is believed that it will not be long before this new generation of encryption technology can be widely applied in society.

### 9.5.2 Scientific issues and key technologies of quantum secure communications

China's overall objective to be reached by the 2050 is to build a practical security communication network based on QKD. The network will be a combination of fiber-optic networks and free-space QKD systems as well as ordinary fiber-optic communication networks. Secure communications that can not be tapped and can not be deciphered will be able to be achieved around the world. The major issues and technologies in need of further research include:

(1) Research on theoretic security; namely, researching how to bridge the inevitable gap between the assumptions of QKD stringent security proofs and the non-ideal existing experiments.

(2) Research on practical security issues. The practical security of QKD has not yet raised too much concern and is a very essential feature of QKD as a tool for information security; hence we should encourage groups involved in information security and privacy research projects to join in the study of how to integrate QKD into the existing secure communication structure.

(3) Development of related technologies. The three elements of information theory for QKD are identification, error correction and private amplification. Identification is the basis of QKD information security capabilities and it is important particularly for the network; efficient error-correcting algorithms closer to the Shannon limit are very critical for QKD as an encryption device and fast private amplification is also quite necessary.

(4) QKD research based on entanglement. This kind of QKD has the characteristic of being more secure than the single-photon scheme, but it is still insufficient both in theoretical analysis and experimental studies.

(5) Development of related devices such as: detectors of communication wave bands, which are fast, efficient, low-dead-time, low-jitter and able to distinguish the number of photons, light sources of fast, high-repetition-rate and narrow-bandwidth single-photons and light sources of entangled photon pairs and so on.

(6) Research on quantum repeaters, which will put remote fiber QKDs to reality along with quantum memory.

(7) Research on network structures; namely, research on how to make best use of QKD to support security and scalable network communications, with particular attention paid to QKD implementation beyond point-to-point

topology structure.

(8) Research on optical communication systems; namely, research of the possibility for QKD to be integrated into confidential optical communications as the encryption support of the physical layer.

(9) Standardization development; that is, to develop and institute the standards and assessment methods of QKD.

### Quantum Secure Communications

Secure communication technology based on quantum key distribution and one-time pad coding can provide non-tapping and indecipherable secure communications in principle.

### Quantum Computing

A new computing model based on quantum mechanical properties (such as superposition and entanglement, etc.) can achieve quantum parallel computing which is different from the classic sequential operations and effectively reduce computing complexity by adopting relevant quantum algorithms.

## References


[1] 2006 NSF Workshop on "Cyber-Physical Systems". 2006. Grand Challenges: Science, Engineering, and Societal Advances Requiring Networking and Information Technology Research and Development, NITRD October 16–17

[2] Clark D. 2009. Architecture from the top down. Spring 2009 FIND Meeting April 6–7

[3] Commission of the European Communities. 2009. Moving the ICT Frontiers—A Strategy for Research on Future and Emerging Technologies in Europe, April 20

[4] Committee on Advancing Software-Intensive Systems Product's Ability, National Research Council. 2008. Preliminary Observations on DoD Software Research Needs and Priorities: A Letter Report. http://www.nap.edu/catalog/12172.html.

[5] DeBenedictis E P. 2004. Will Moore's law be sufficient? //Proceedings of the ACM/IEEE Conference on Supercomputing, Nov

[6] European Commission, Information Society and Media, Internet of Things in 2020- Roadmap for the Future.Version 1.1, 2008

[7] Fisher D. 2007. US National Science Foundation and the future Internet design. ACM SIGCOMM Computer Communication Review, 37(3)

[8] GENI Planning Group. 2006. GENI design principles. IEEE Computer, 39(9)

[9] Goldin D, Wegner P. 2008. The interactive nature of computing: Refuting the strong Church-Turing thesis. Minds and Machines, 18(1)

[10] Hirvonen J, et al. 2007. Sensor Networks Roadmap. http://www.vtt.fi /inf/pdf/tiedotteet/2007/T2381.pdf

[11] Hughes R, et al. 2004. A Quantum Information Science and Technology Roadmap, April 2

[12] Institute for the Future, 2005–2055 Science & Technology Outlook: Perspectives [SR-967], May 2006

[13] International Energy Agency. 2008. Energy Technology Perspectives 2008: Scenarios & Strategies to 2050

[14] International Technology Roadmap for Semiconductors (ITRS)2007(2008 revision). http://www.itrs.net/about.html

[15] Jackson D, et al. 2008. Software for Dependable Systems: Sufficient Evidence? Committee on Certifiably Dependable Software Systems, National Research Council. http://books.nap.edu/catalog/11923.html

[16] Kamps M D, Knolla A C. 2007. Roadmap for neuroIT: Challenges for the next decade. IEEE Engineering in Medicine and Biology Magazine, May/June

[17] Karp R M. 2008. Understanding science through the lens of computation. ICDCS, June 17

[18] Kramer B. 2006. Software Roadmap to Plug and Play Petaflops, July

[19] Marron P J, et al. 2006. Embedded WiSeNts Research Roadmap, Embedded WiSeNts Consortium, Nov

[20] National Science Foundation Cyberinfrastructure Council, Cyberinfrastructure Vision for 21st Century Discovery, March 2007

[21] Perry W, Signori D. 2001. A Mathmatical Framework for Measuring for the Effects of Information and Collaboration on Shared Awareness, 6th ICCRTS

[22] Plummer J D. 2001. Material and process limits in silicon VLSI technology// Proceedings of IEEE, 89(3)

[23] President's Council of Advisors on Science and Technology. 2007. Leadership under Challenge: Information Technology R&D in a Competitive World, August
[24] President's Information Technology Advisory Committee. 2005. Computational Science: Ensuring America's Competitiveness, June
[25] Purdy J G. 2003. The next 50 years in mobile and wireless. Mobile Imperative,1
[26] Roco M C, Bainbridge W S. 2002. Converging Technologies for Improving Human Performance: Nanotechnology, Biotechnology, Information Technology, and Cognitive Science. NSF/DOC-Sponsored Report
[27] Stevens R. 2006. CTWatch quaterly. Trends and Tools in Bioinformatics and Computational Biology, 2(3)
[28] Swan M. 2008. The future of technology. Technology and Society Committee, Nov 11
[29] Takuo Imagawa. 2005. Japan's Policy Initiatives toward Ubiquitous Network Societies, April 7. http://www.itu.int/osg/spu/ni/ubiquitous/Presentations/5_imagawa_japan.pdf
[30] Tschudi W, et al. 2004. Roadmap for Public Interest Research for High Performance Data Centers. Lawrence Berkeley National Lab, March 30
[31] Tummala R R. 2004. SOP: What is it and why? A new microsystem integration technology paradigm—Moore's law for system integration of miniaturized convergent systems of the next decade. IEEE Trans on Advanced Packaging, 27(2):241–249
[32] Watanable H. 2007. Advanced CMOS and related characterization. International Conference on Frontiers of Characterization and Metrology for Nanoelectronics
[33] Wing J M. 2006. Computational thinking. Communications of ACM, 49(3): 33–35

# Epilogue

Not only it is the first time for us to conduct a strategic research for a roadmap for the development of S&T in 40 years, but it is also rare in the world. There are great differences between this long-term roadmap and general roadmaps, and there is no ready experience for reference, so we can only learn in practice. We feel that, for the roadmap of a long period, we can not emphasize the implementation and operational issues like those in short-term roadmaps, but must pay more attention to identify the strategic requirements and the development directions. We have tried to integrate the economic and social development trends and strategic requirements of China by 2050 into the research roadmap. By 2050, China may become the world's largest market in IT sector, and the IT development must benefit the entire population and take the road of sustainable development. As China gets closer to the international frontier in IT, its IT sector must gradually become an important member of the international innovation community.

Through strategic research of near two year, we have made a basic observation: IT is quite different from the mechanical and electrical technologies. After half a century of rapid development, it has not become a traditional industrial technology mainly based on the incremental improvement. It is facing a new revolution in the whole $21^{st}$ century. IS&T will be interweaved with biology, nanotechnology, cognition, and other technologies, continue to show its vigor to lead and support national economic development and change the way of people's life.

Integrated circuits, high-performance computers, the Internet, and memory and storage, will encounter unprecedented obstacles from 2020 to 2030 if we only rely on the continuation of existing technologies. This indicates the need for principle-based scientific discoveries and technological breakthroughs. Our knowledge level and foresight are limited, so we can not really predict exactly what the mainstream technology shall be in such fields as chips, network and computers in the year 2050 or 2035. Currently we can only come to the following conclusions: before 2020, we shall actively explore the core technologies for breaking "the wall of IT", focusing on the solution of such problems as scalability, low power consumption, security and ease of use of the information systems; it shall be gradually clarified which technology will become the new mainstream technology after 2020; there will be great revolutions in IT between 2020 and 2035; the ubiquitous information network over the world based on breakthrough technologies will be gradually formed from 2035 to 2050.

Since 1940s, the third technological revolution symbolized by IT has been going on for 70 years, and the world financial crisis in the early 21st century essentially indicates that technology innovations in IT area are not strong enough to drive the economical development. The next 20 years shall be a period to produce IS&T breakthroughs in the next economic long wave through innovative changes. At present, China has not yet focused on forward-looking research projects such as nano devices, the Post-IP Internet, and Exascale supercomputers, it might miss the fleeting opportunities.

Under the unified organization of Chinese Academy of Sciences, we have studied for almost two years on long-term development strategy and completed a task we have never done before. Our judgment may be fraught with improperness, and some words may be inaccurate or unclear, so we welcome corrections from readers. In the process of doing strategic roadmap research, we have got the guidance and support from many leaders and experts as well as scholars and we hereby present our sincere gratitude to them.

Strategic Study Group on IS&T of the

Chinese Academy of Sciences

October, 2010

Kolibri 360 GmbH
Ferdinand-Jühlke-Straße 7
99095 Erfurt, Deutschland
produktsicherheit@kolibri360.de